The Archaeologist's Manual for Conservation

THE ARCHAEOLOGIST'S MANUAL FOR CONSERVATION

A Guide to Non-Toxic, Minimal Intervention Artifact Stablization

Bradley A. Rodgers

Program in Maritime Studies
East Carolina University
Greenville, North Carolina

KLUWER ACADEMIC / PLENUM PUBLISHERS
New York, Boston, Dordrechty, London, Moscow

Library of Congress Cataloging-in-Publication Data

Rodgers, Bradley A.
The archaeologist's manual for conservation / Bradley A. Rodgers.
p. cm.
Includes bibliographical references and index.
ISBN 0-306-48466-8 – ISBN 0-306-48467-6 (pbk.)
1. Antiquities–Collection and preservation–Handbooks, manuals, etc. I. Title.
CC135.R64 2004
930.1′028–dc22 2004041844

ISBN 0-306-48466-8 (hardbound)
0-306-48467-8 (paperback)

233 Spring Street, New York, New York 10013

http://www.wkap.nl

10 9 8 7 6 5 4 3 2 1

A C.I.P record for this book is available from the Library of Congress.

Printed in the United States of America.

To my wife Annie—through thick and thin

Foreword

This is a Foreword by an archaeologist, not a conservator, but as Brad Rodgers says, "Conservation has been steadily pulled from archaeology by the forces of specialization" (p. 3), and he wants to remedy that situation through this manual. He sees this work as a "call to action for the non-professional conservator," permitting "curators, conservators, and archaeologists to identify artifacts that need professional attention and, allow these professionals to stabilize most artifacts in their own laboratories with minimal intervention, using simple non-toxic procedures" (p. 5). It is the mission of Brad's manual to "bring conservation back into archaeology" (p. 6). The degree of success of that goal depends on the degree to which archaeologists pay attention to, and put to use, what Brad has to say, because as he says, "The conservationist/archaeologist is responsible to make preparation for an artifact's care even before it is excavated and after its storage into the foreseeable future"... a tremendous responsibility" (p. 10).

The manual is a combination of highly technical as well as common sense methods of conserving wood, iron and other metals, ceramics, glass and stone, organics and composits — a far better guide to artifact conservation than was available to me when I first faced that archaeological challenge at colonial Brunswick Town, North Carolina in 1958—a challenge still being faced by archaeologists today.

The stage of conservation in 1958 is in dramatic contrast to the procedures Brad describes in this manual—conservation has indeed made great progress. For instance, a common procedure then was to heat the artifacts red hot in a furnace—a method that made me cringe. Faced with the need to treat thousands of wrought nails I was excavating from the ruins of Brunswick Town, I rented a sandblaster and was able to clean the rust from padlocks, hinges, pintles, keys, buckles and tools, as well as wrought nails, at a cost of 25 cents each (South 1962a:18), saying, "No other method can match this achievement." I also said, "No soaking to remove salts has been found to be necessary,"— a statement that probably makes Brad cringe today. Later I soaked a bucket of nails in a diluted 5% solution of muriatic acid to remove the rust, decreasing dramatically the cost per nail.

In 1962, I sent to my supervisor, "Notes on Treatment Methods for the Preservation of Iron and Wooden Objects" (South 1962b), in which I described the treatment of water-soaked wood and iron recommended by various conservators and archaeologists at the time—such as soaking cannonballs in a fresh water stream to remove the salts. Artifact conservation was a process being tried using various nefarious, and some efficacious, methods that archaeologists were pulling out of the seat of their pants.

In later years my experience with conservators was not the best as that "force of specialization" Brad speaks of widened the gap between professional conservators and the conservation needs I had as an archaeologist. For instance, when a professional conservator was made available to me, I gave 10 nails to the professional to process for use in an exhibit—three months later four were done. I received a promise that the others would be completed within a year—these were not the crown jewels of Europe! Well, back to sandblasting and muriatic acid.

Since that time, I have left all Spanish barrels I discover (after recording and photographing them *in situ*) in the well hole where they have been safely conserved by nature for four and a half centuries. As Brad says: "*In-Situ* conservation is designed to extend an artifact's stability in the ground or the sea bottom, a form of preemptive conservation" (p. 10) — a wise method until such a time that conservators can successfully conserve composite artifacts.

I could continue, but the point is many archaeologists are very likely still trying to conserve their artifacts by using methods pulled from the seat of the pants, because the sources written by conservators (and there is a definitive listing of them in Brad's manual) are too complex and technical for use by field archaeologists — yet the responsibility to conserve the artifacts is constantly there.

Brad's manual in this volume is the first attempt to "bring conservation back into archaeology." It is my hope that this manual will help do that, so further horror stories can be reduced and archaeologists can come closer to achieving "Pinky" Harrington's goal—"To preserve the physical remains of our past and to employ them in perpetuating our historical heritage" (Harrington 1965:8, quoted in South 1976:42).

Stanley South H. H. D.
The University of South Carolina
South Carolina Institute of Archaeology and Anthropology
Columbia, South Carolina

REFERENCES

Harrington, J. C. 1965. *Archaeology and the Historical Society.* The American Association for State and Local History. Nashville.

South, Stanley 1962a. A Method of Cleaning Iron Artifacts. *Newsletter of the Southeastern Archaeological Conference* 9(1): 17–18.

—. 1962b. Notes on Treatment Methods for the Preservation of Iron and Wooden Objects (Unpublished manuscript, North Carolina Department of Archives and History, Brunswick Town State Historic Site, Raleigh.

—. 1976b [1972]. The Role of the Archaeologist in the Conservation-Preservation Process. Pp. 35–43. In *Preservation and Conservation Principles and Practices.* Edited by Sharon Timmons. Produced by the Smithsonian Institution Press. Published by the National Trust for Historic Preservation in the United States. The Preservation Press. Washington, D.C.

Preface

This project began over a decade ago in my first attempt to produce a manual that outlined the procedures used at the artifact conservation laboratory at East Carolina University to stabilize and conserve water-degraded archaeological artifacts. The *Conservator's Cookbook* (1992), became a practical conservation manual designed to compliment graduate level conservation classes. The text and flow charts provided simple hands-on advice to students who wanted to conserve artifacts. Complimenting the conservation manual was the *Conservation of Water Soaked Materials Bibliography* (1992), by far the most comprehensive listing of books and articles yet compiled concerning archaeological artifact degradation, conservation, and stabilization. These texts enjoyed surprising popularity, inspired no doubt by the fact that they endorsed practical application and methodology in the midst of a field whose literature was dominated by theory with anecdotes from spectacularly difficult ostentatious projects.

At that time, however, crossover students from the Department of Anthropology began to add great scope to my experience, expanding significantly the types and conditions of artifacts brought into the laboratory for conservation. I soon discovered that although the treatments in the *Conservator's Cookbook* were specifically designed to stabilize water soaked and degraded artifacts, they worked equally well or better on artifacts recovered from land sites. Indeed, the procedures devised for water-degraded material, reflected a considerable advance over those employed to stabilize artifacts in most terrestrial laboratories.

Experience proved that the famed preservation properties of seawater are simply an anomaly of available conservation technology. In other words, given the same treatments, artifacts excavated from the ground demonstrate equal or better preservation to those recovered from wet sites. It also became clear that dry recovered artifacts can be conserved in a condition that permits detailed micro-excavation and examination of both prehistoric and historic objects. Native American copper ornaments from Fort Neoheroka (1713) could be examined for signs of use and polishing. What was thought to be a concreted gun barrel from Santa Elena (1572) was found to be no more than a mandrel rolled piece of sheet iron, and buttons and coins lost all concretion and accumulated filth to reveal dates, maker's stamps, gilding, mold marks, and wear that was never seen before.

In light of these discoveries it became increasingly clear, that despite the fact that underwater archaeology is a much younger field than prehistoric and historic archaeology and owes much to its parent fields, it is now well ahead in artifact conservation technique and practice. *The Archaeologist's Manual for Conservation* is the first attempt to share these hard won conservation advances with the archaeological field as a whole, and demonstrate for the first time that hands-on conservation should be endemic to the entire profession; prehistoric, historic, and underwater.

Acknowledgements

Every author is in debt to innumerable people who share ideas, concepts, and opinions, but most of all, to those who offer encouragement. Without a first step, a book remains only a pipe dream, and few first steps are taken without encouragement. In this light I would like to thank my colleague Dr. Annalies Corbin, for her ideas, help in proofing, and sage advice to the effect of, "what are you waiting for"? She and Dr. Nathan Richards shouldered heavy burdens as I worked, proof positive that "old fashioned teamwork" is, without question, still alive and well. Dr. Richards also provided fine detailed line drawings that elucidate conservation concepts far beyond the means of written text. I would also like to thank my graduate help, Mr. Dave Krop, Mr. Andrew Weir, Ms. Claire Dappert, and especially Ms. Dannielle La Fleur, your diligence and hard work are much appreciated. Special thanks go to Mr. Christopher F. Valvano for the photographs, each a major endeavor.

Scientists can sometimes focus their entire attention on concept detail—learning a lot, about a little. Conservation science is the opposite. An archaeologist/conservator needs to know a little, about a lot. It is good to have friends, therefore, who can give sound and specialized advice. In this light I would like to thank Mrs. Marcia Kuehl, chemistry consultant and old friend and Dr. Charles Ewen for making available the collections at the Phelps Anthropology Laboratory for study and preservation.

Final thanks go to my wife Ann for your encouragement and patience. Rusted log staples don't belong on the kitchen table and I won't tell you what has been in the freezer.

Contents

List of Figures

Chapter 4

Chapter 5

Chapter 6

Chapter 7

Chapter 8

Introduction

Conservation Is Part of Archaeology

This manual is designed to take the mysticism out of archaeological artifact conservation and act as both reference and guide. It is intended to be a tool to assist archaeologists in stabilizing a majority of the artifacts they excavate, or those already in storage. These stabilized archaeological collections will be preserved into the future, permitting reexamination and multiple interpretations of the data as our knowledge base grows through time. In addition, conservation will permit improved in-depth primary artifact interpretation, as fully conserved artifacts reveal fabrication, wear patterns, and detail impossible to detect in non-conserved artifacts. Conservation, therefore, is a critical tool within archaeology, a tool that becomes less meaningful if it is isolated, or seen as merely a technical skill that can be farmed out to the "hard sciences." *The Archaeologist's Manual for Conservation* is intended as a counterpoint to the popular specialization trend. My goal in offering this manual is to put artifacts back in the hands of archaeologists or material culture specialists who can best decipher them, opening avenues of artifact or material culture interpretation that are disappearing as artifacts either decay in storage or are sent away to the "conservation professionals."

Yet it cannot be denied that artifact conservation has become an increasingly complex subject. The manual, therefore, is divided into chapters of material types and subdivided into easily understood components within each chapter, each backed by extensive bibliographic reference. Of the hundreds of treatment concepts available today, the few promoted in the manual are chosen for their effectiveness, simplicity, lack of toxic effects (on both conservator and environment), and cost effectiveness. All treatments must also conform to the rules of conservation specified in Chapter 1.

The *Archaeologist's Manual for Conservation* can be approached on several levels. It can be read in narrative form for those interested in a particular material or conservation procedure, or used as a practical reference by following the artifact treatment flow charts on the first page of each chapter. Chapters open with material theory and take the reader through first, a short history of a material followed by a theoretical description, including the refining process if the material is a mineral, or the biology of the substance if it is an organic. Theory sections will also describe the decomposition and deterioration of an artifact. The practical side of the manual follows the theoretical sections. Methodology explains in detail how any given material type is stabilized.

HOW THE MANUAL WORKS AND ITS ANTECEDENTS

The Archaeologist's Manual for Conservation outlines practical methodology in simple flow chart form. The flow charts demonstrate how an artifact can be stabilized from start to finish and the reasoning behind the use of the method deemed most effective. The flow charts are located on the first page of each chapter concerning a particular substance such as wood, iron, copper, glass, ceramic, and so on. Treatment simply follows the logic of the flow charts. Since the actual explanation for flow chart procedures is fairly complex, and much too detailed to fit the confines of the diagram, each procedure in the chart refers to the pages in the text where that treatment is explained. The methodological explanations also reflect insights on similar or related materials or procedures.

Identifying the substance that makes up an artifact is equivalent to finding the ON switch to this manual. It allows a technician to turn directly to the flow chart guiding treatment for that material. Though material make-up is obvious in most instances, a technician's ability to identify the elements that make up an object may at times be difficult. Strategically this task is of paramount importance, as it forms the basis for all subsequent treatment.

At times it is not good enough to simply identify that an artifact is made out of wood, ceramic, or iron. Treatment often depends on the particular type of wood, ceramic, or iron. Archaeologists and curators trained in material culture and conservation procedures, enjoy a distinct advantage in this area over non-archaeological conservators. Their experience permits them to determine artifact and material type at a glance, despite its present condition or accumulated concretion. A scupper liner for example will always be lead, a nail will always be wrought iron, and a cannon ball will almost always be cast iron. Conservators not trained in archaeology or artifact identification are often forced to rely on material analysis alone to confirm material type, adding time and cost to the conservation procedure. Artifact typing argues powerfully for a continuous dialogue between conservators, curators, and archaeologists if not cross training in these professions.

The Archaeologist's Manual addresses material typology with simple diagnostic methods and a list of traditional artifacts fabricated of that material at the beginning of each chapter. These identification methods are usually effective in separating materials precluding the use of more sophisticated and costly methods of identification such as X-ray diffraction analysis, PIXI, and gas chromatography. However, these procedures do have their uses and can be found in most physics departments in local universities. Textile and organic analysis can also be accomplished in a Home Economics or Biology Department. Networking is encouraged in conservation and is often the key to promoting good and complete analysis of an artifact.

This manual will also list most of the other conservation methods currently available for treatment of a certain sort of material in chart form following the treatment flow chart. The treatments compilation chart will not go into detail concerning how these other methods are used, only that under certain circumstances, they are, or have been in the past. This simple list of conservation methods is augmented by a chapter by chapter bibliography of up to date research in the various journals and publications concerning conservation of artifacts. Therefore, each section of material type is serviced with a listing of the latest sources. If an article or book covers more than one type of material it is included in the bibliographies of each of the material types mentioned. General conservation works are included after Chapter 1, "The Minimal Intervention Laboratory." Research for this manual includes over 1000 books and articles gleaned from over 100 journals in conservation, archaeology and related fields in engineering, biology, wood science, chemistry, biochemistry, as well as various curatorial journals (see appendix A).

Conservation is a relatively new field within archaeology and few books specifically address archaeological conservation. Most of the classic references in conservation are written from the point of view of a museum curator. These include Plenderlith and Werner's, *The Conservation of Antiquities and Works of Art*, first published in 1956, and Muhlethaler and Barkman's *Conservation of Waterlogged Wood and Wet Leather*, 1973. More recent books concerning archaeological conservation include, Colin Pearson, ed., *Conservation of Marine Archaeological Objects*, 1987; J. M. Cronyn's, *The Elements of Archaeological Conservation*, 1990; Donny Hamilton's, *Basic Methods of Conserving Underwater Archaeological Material Culture*, 1996, as well as specialized works such as Wayne C. Smith's, *Archaeological Conservation Using Polymers*, 2003. Though excellent sources of information in their own right, they tend to be written from a theoretical perspective and are generally speaking, written for conservators, conservation scientists, or archaeologists with substantial backgrounds in artifact stabilization theory. In essence, they do not completely cross the practical application gap, and in fact, tend to create an artificial separation between conservation and archaeology.

At times, however, practical research has surfaced. Works by Wendy Robinson, *First Aid for Marine Finds*, 1981, followed by Catherine Sease's, *A Conservation Manual for the Field Archaeologist*, 1987, and Katherine Singley's, *The Conservation of Archaeological Artifacts From Freshwater Environments*, 1988, attempted to bring conservation back within the realm of practical application and archaeology and succeeded in doing so to some degree. Unfortunately, a more comprehensive fully practical manual did not immediately follow their lead and conservation has been steadily pulled from archaeology by the forces of specialization.

CONSERVATION AND ARCHAEOLOGY

Colin Pearson states in the Preface of Butterworth's, *Conservation of Marine Archaeological Objects* that, "excavation without conservation is vandalism (Pearson, 1987: Preface)". It should be stated plainly here that this is true not only for artifacts recovered from underwater sites, but artifacts recovered from terrestrial excavations as well. Yet, one of the great myths in terrestrial archaeology is that artifacts are safe and stable simply if they are cleaned of surface debris, tagged, and stored. In reality, the deterioration process and chemical breakdown of artifacts recovered from a marine environment, a freshwater or wet site environment, or a terrestrial environment, will continue even after drying and storage. This is particularly true of metals and organics. The difference seems to be the rate at which artifacts disintegrate.

Generally terrestrial and fresh water or wet site recovered artifacts deteriorate at a somewhat slower pace than artifacts recovered from the sea, even though objects recovered from the ocean may initially look like they are in better shape. But deterioration is inevitable in both land and water recovered artifacts. Material recovered from any of these physically different sites cannot always be effectively stabilized with simple exterior cleaning and storage and short term observations (several years) are not always a reliable indicator of whether an artifact is stable. Curation has no time limit, artifacts are expected to last as long, for all practical purposes, as our civilization does. Therefore, conservation is a necessity for long-term artifact storage, without it an object's life span is greatly reduced.

The reasons for artifact deterioration are complex and vary with each artifact, but stated simply, in most instances an archaeological artifact has reached a natural chemical and electrical equilibrium with its environment, whether with the ground or the ocean benthic environment. When that equilibrium is disturbed through excavation, the artifact will begin to break down. Since artifacts recovered from the ocean break down at a spectacular rate, underwater archaeologists have been forced to devise techniques to counteract artifact deterioration. The old standard practice of most archaeological laboratories, exterior cleaning and storage, had no appreciable affect in stabilizing water-degraded artifacts.

The tagged, bagged, boxed, and shelved practice is an alarming yet unquestioned *modus operandi* for most archaeological laboratories; alarming because the artifacts so processed are NOT necessarily stable and may deteriorate over time even in air tight containers. Years after the artifacts are put away for "safe keeping," when the contents of the storage containers are examined (or rediscovered as it were), one of three observations will invariably take place. First, if the artifact is robust (lithic, bone, glass, or ceramic) it may appear that it is normal, or perhaps it is a slightly different hue, color, or iridescence than its accession photograph. The curator is left to decide if the photo or the artifact is changing color, but no great injury or harm seems to have been done to the artifact. Second, a more friable sort of artifact (metallic, textile, cordage, organic, or wood) has over time become several artifacts while untouched in its storage box. This artifact is actively spalling, breaking up, and or leaving residue. In most laboratories this is overlooked, seen as normal aging, or perhaps the result of a storage problem. In the third scenario the curator opens the artifact storage box and is left to guess where the artifact went, and who replaced it with a pile of dust and debris? These scenarios are obviously a temporal snapshot of what will happen to many artifacts in storage (even some of the more robust specimens) that have not undergone conservation, and these scenarios are repeated at virtually every archaeological laboratory in the country.

The Archaeologist's Manual for Conservation is intended to allow archaeologists and curators of archaeological artifacts to examine their collections in a whole new light, permitting these professionals to view the effects of time on unstabilized artifacts for what it really is, a destructive process with only one eventual outcome. This manual can be seen as a call to action for the non-professional conservator. It will permit curators, conservators, and archaeologists to identify artifacts that need professional attention and, allow these professionals to stabilize most artifacts in their own laboratories with minimal intervention, using simple non-toxic procedures.

In this way *The Archaeologist's Manual* is somewhat iconoclastic and works to dispel the notion that all archaeological conservation is too difficult and complex for the average curator or archaeologist. It will remove or explain in simple terms professional conservation language. For although professional language promotes easy communication within a vocation, it intimidates and distances people outside a profession. For so long as curators and archaeologists buy into the notion that conservation is too complex and difficult a subject to tackle on their own (on any level), the collections will continue to languish under inaction and neglect. The final irony remains that while many objects could be saved with very simple procedures, they may in fact be doomed while awaiting professional attention, professional attention that may simply be too costly to be practical. Few archaeology labs can hire professional conservators, and fewer still can afford to ship their entire collections to those laboratories that can stabilize them.

Though *The Archaeologist's Manual for Conservation* suggests that curators and archaeologist can take care of some of their conservation needs, it does not intend to imply that neophyte conservators attempt to stabilize large complex or complicated composite artifacts. Overly complex projects concerning valuable artifacts or works of art should be left to teams of professional conservators working with the latest scientific equipment and knowledge of the subject. A general rule of thumb for this manual is, if a reader has concerns about an artifact, or is unsure of their ability to stabilize an object without damage; it should go to a professional or a professional should be consulted. Conservation can be a very complex subject and is governed by how highly degraded an object may be, the type of degradation it has suffered, the rules of conservation, goals of treatment, laboratory philosophy, and laboratory type. The size and scope of this manual is manageable only by assuming that the reader here is concerned with the least complex of conservation goals (stabilization), the simplest laboratory type, and the least intrusive non-toxic, safe procedures. Complex and specialized procedures are included within written research in the chapter bibliographies, but are not considered relevant to the scope or design of this book.

In the final analysis *The Archaeologist's Manual for Conservation* is an attempt to make archaeologists and archaeological curators aware that artifacts are not just a valuable resource recovered during an excavation, but can continue to shed light on a site long after excavation and recovery. But only if the recovered objects are preserved into the foreseeable future. In practical application, conservation techniques and micro-excavation analysis can add a great deal to the interpretation of any site. Conserved artifacts more easily reveal adaptively reutilized artifacts, tool patterning, manufacturing techniques, wear patterns, stains, gilding, makers marks, mold marks, nicks and dents from use. However, this manual is simply a tool. The archaeologist/ conservation technician must make it work and bring conservation back into archaeology.

INTRODUCTION: CONSERVATION ETHICS SOURCES

Anon. "A Code of Ethics for Conservators." *Museum News* 58:4 (1980): 28–34. [ECU-109]

Appelbaum, B., and P. Himmelstein. "Planning for a Conservation Survey." *Museum News* 64:3 (1986): 12–14. [ECU-226]

Arnell, Jack. "The Importance of Conservation Must be Recognized." *Bermuda Maritime Museum Quarterly* II (1989): 22. [ECU-12]

Ashley-Smith, J. "The Ethics of Conservation." *The Conservator* 6 (1982): 1–5. [ECU-412]

Croome, A. "Boost for Marine Archaeology." *New Scientist* 50:257 (1971): 746–747. [ECU-112]

Hamilton, Donny L. "Conservation in Nautical Archaeology." *Proceedings* (12th Conference on Underwater Archaeology, 1981). [ECU-39]

Leigh, D. "Reasons for the Preservation and Methods of Conservation." *Marine Archaeology.* D.J. Blackman, ed., Anchor Books, 1973: 203–218.

Lodewijks, I.J. "Ethics and Aesthetics in Relation to the Conservation and Restoration of Waterlogged Wooden Shipwrecks." *Conservation of Waterlogged Wood* (1979): 107–110. [ECU-313]

Logan, Judith A. "Thoughts on the Role of the Archaeological Conservator." *CCI Newsletter* (Canadian Conservation Institute) (1988): 810. [ECU-51]

Organ, R.M. "A Conservation Institute: A Personal Concept." *Journal of the American Institute for Conservation* 2 (1975): 63–65.

Pearson, C. "Legislation for the Protection of Shipwrecks in Western Australia." *ICOM Proceedings* (1975).

Peterson, Curtiss E. "The Role of Conservation in the Recovery of the USS *Monitor*." *Proceedings* (15th Conference on Underwater Archaeology) (1984): 195–196. [ECU-75]

Rodgers, Bradley A. "Introduction." *The East Carolina Conservator's Cookbook: A Methodological Approach to the Conservation of Water Soaked Artifacts.* Herbert R. Paschal Memorial Fund Publication, East Carolina University, Program in Maritime History and Underwater Research, 1992. [ECU-402]

Tighe, Mary Anne. "The Right Restorer." *Washingtonian* (February 1972): 121–122. [ECU-393]

Ullberg, A.D. and R.C. Lind, Jr. "Consider the Liability of Failing to Conserve Collections." *Museum News* (1989): 32–33. [ECU-1131

Chapter 1

The Minimal Intervention Laboratory

ARCHAEOLOGICAL CONSERVATION

Artifact conservation has been defined in several different ways depending on the professional viewpoint of the author. Archaeological artifact conservation for instance, can be defined in a somewhat more restrictive manner than art or building conservation, that usually includes phrasing concerned with restoration. Unlike historic site curators or professional resource managers, archaeologists are not generally concerned with restoring an artifact to anything approaching its original appearance, and in fact, can tell more about an object if it is worn, well used, and broken, than if it is in pristine condition. While archaeological conservation places no premium on restoring an artifact's appearance, it does become a more inclusive profession in other respects. Lindstrom and Rees-Jones define marine archaeological conservation in two ways stating that the goals of conservation are "the control of the environment to minimize decay" and secondly, "to stabilize them [artifacts] where possible against further deterioration (Pearson, 1987; Series Editor's preface)." The multi-component approach to archaeological conservation is a point well taken and over the years it has become more encompassing.

Since this definition was postulated, the separation between archaeologist, and archaeological conservator and, curator has become, if anything, more blurred. Job distinctions now depend on where the skills of a crewmember fit into the overall archaeological team and how the team is set up in the planning for fieldwork. Archaeological conservators and conservation technicians don't just work in laboratories anymore than archaeologists only work in the field. It becomes even more convenient if they are one and the same person.

Nonetheless, the new conservator/ archaeologist actually begins the conservation process during the initial project research design, when it is planned that artifacts will, or perhaps will not (as in the case of a pre-disturbance survey), be recovered. Current interest in cultural resource management (CRM) emphasizes that artifacts be removed from a site only under the best planned circumstances, and only if certain conditions are met. Namely that conservation funding be set in place before the excavation, or a fully capable laboratory and staff support the field work while museum, state, or academic facilities agree to store and curate the artifacts. As archaeological conservation becomes more widespread and implemented, artifacts can even receive *in-situ* stabilization treatment and be left in place. *In-Situ* conservation is designed to extend an artifact's stability in the ground or the sea bottom, a form of preemptive conservation.

Therefore, in a modern sense, archaeological artifact conservation starts well before any artifact is excavated and continues well after an artifact is curated. The definition of conservation has to be inclusive and the conservator by necessity wears many hats. For the purposes of this text archaeological artifact conservation is; the designed implementation of processes, procedures, and strategies to guard artifacts and archaeologically important objects, as far as possible, against deterioration on site, during recovery, while transitioning to storage, or while in storage or under curation and or exhibit. Conservation in this sense is a verb that also includes the micro-excavation and examination of artifacts as they undergo treatment, in an effort to gain information concerning artifact fabrication, site formation process, use, and wear. This definition is far reaching and temporally has no boundaries. The conservator/archaeologist is responsible to make preparation for an artifact's care even before it is excavated and after its storage into the foreseeable future. It is no coincidence that as more archaeologists are trained in both conservation and cultural resource management—that fewer and fewer artifacts are being recovered from archaeological sites, they represent a tremendous responsibility. After all, the object of archaeology is the collection of data and information to add to our knowledge base of the human past, full excavation with recovery of every artifact at a site may be unnecessary and comes with a large price, curation and conservation, in the end, are forever.

COMPREHENSIVE LABORATORIES

Conservation facilities supporting archaeological work range in size and complexity. There is no general principle concerning laboratory size only general principles regarding design for intended use. For the sake of organization most artifact conservation facilities can be divided into three categories, large comprehensive purpose built facilities, mid-sized complete service facilities (that have generally grown through contract work from program oriented labs), and program supporting laboratories (the special concern of this manual). Laboratory size, in square footage, is not necessarily an indicator of how well a laboratory can perform, nor is budget or equipment. Since archaeological conservation is extremely labor intensive, a far better indicator of lab performance is the knowledge, expertise, and experience of the conservator and technicians working at a lab.

There are few large purpose built archaeological conservation laboratories in the world. Comprehensive laboratories operate on large budgets, are university, government, community, contract and grant supported. They may employ scores of conservation scientists and technicians and originally were constructed from the need to support large, state, university, or national projects. Examples of comprehensive conservation facilities include the Canadian Conservation Institute, the

Maryland Materials Laboratory, the Texas A & M Laboratory, the CSS Hunley Laboratory, the Mary Rose Laboratory, the Vasa Laboratory in Stockholm, the Skuldelev Laboratory in Denmark, and the Western Australia Maritime Museum Conservation Laboratory. Over the years Comprehensive labs have produced and pioneered the techniques that have laid the foundation for advances in the field of marine archaeological conservation.

However, since the majority of these laboratories were built to conserve artifacts from shipwreck sites there has been a disconnection in the transfer of advanced conservation technology to terrestrial archaeology laboratories. The reasons for this divide are understandable. The techniques used and pioneered in the large comprehensive facilities were intended for water-degraded materials. The technical relevance of using marine techniques for dry land excavated materials was a largely unexplored question, answered through the experience of smaller duel purpose facilities. Mid-sized, complete service laboratories picked up these techniques and began to use them for both land and sea recovered artifacts. The duel use of these conservation techniques has since proved that the techniques originally intended to stabilize water degraded artifacts work extremely well on objects recovered on terrestrial sites.

COMPLETE SERVICE LABORATORIES

Mid-size Complete Service Laboratories bridge the gap between the large comprehensive labs and simple program supporting archaeology laboratories known for our purposes as Minimal Intervention Laboratories. Mid-sized complete service labs pass on this knowledge first by researching different conservation methods for their own use and then using and passing on the most efficient, cost effective conservation techniques developed by the large Comprehensive Labs. Generally speaking, Complete Service Labs have evolved from small program supporting university labs by using contracts and grants to acquire and build some of the specialized equipment needed. Complete service facilities are often called on to handle virtually any conservation problem from terrestrial or underwater sites yet their budgets and facilities are not up to the standards of the large comprehensive facilities. Complete service laboratories must rely on only the most cost effective dependable conservation techniques, since they cannot afford the expense of redundant procedures, equipment, and chemical storage. By-and-large the use of toxic chemicals and questionably dangerous procedures has also been eliminated at these laboratories for the safety of the student technicians working there, and to cut costs in dealing with expensive, perhaps redundant, equipment and difficult to handle and dispose of chemicals. Full service labs spend a good deal of time in researching conservation procedures that fit within the parameters of the laboratory design, philosophy, and budget. A mid-size complete service lab, as its name implies, can conserve virtually any artifact, but cannot offer the variety of conservation techniques presented by the comprehensive facilities.

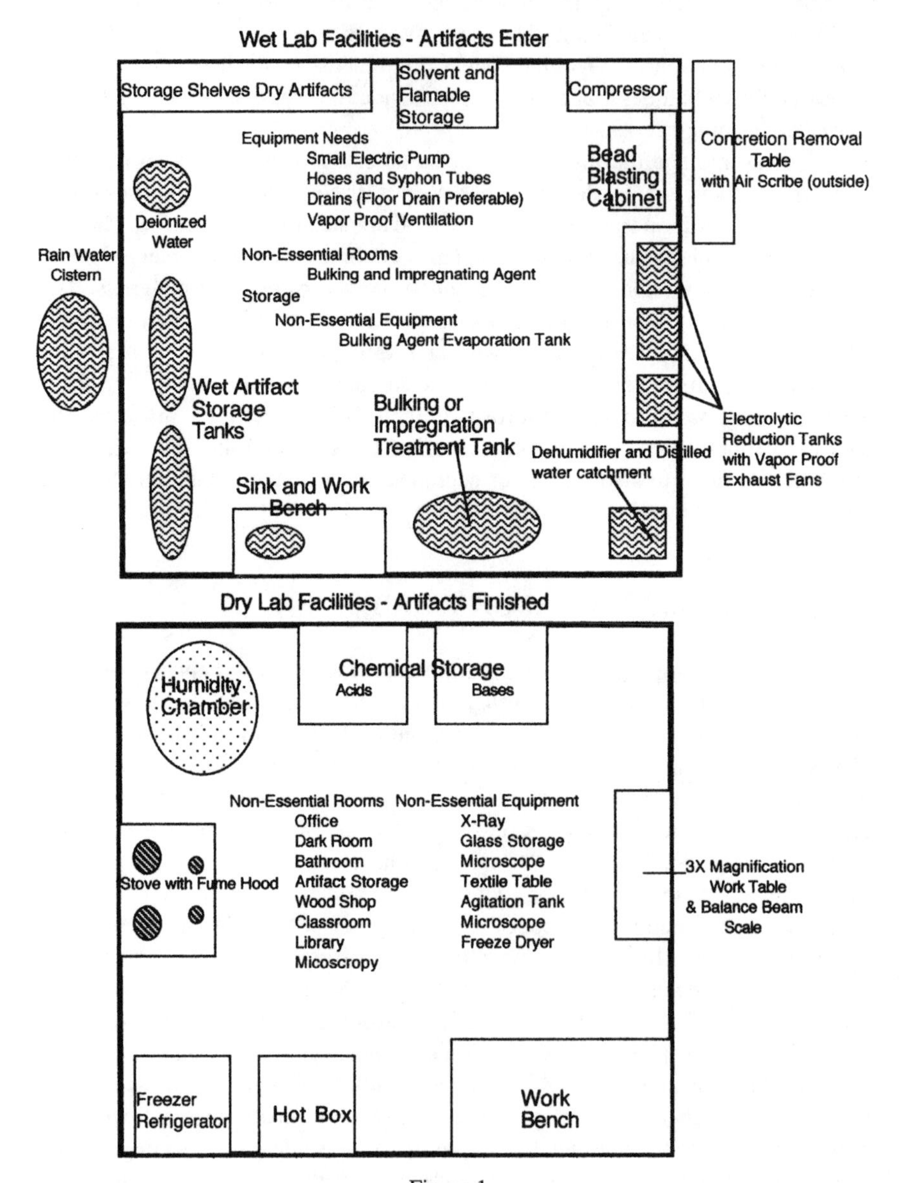
A Minimal Intervention Artifact Conservation Laboratory
Wet Lab Facilities - Artifacts Enter
Storage Shelves Dry Artifacts
Solvent and Flamable Storage
Compressor
Concretion Removal Table with Air Scribe (outside)
Bead Blasting Cabinet
Equipment Needs
Small Electric Pump
Hoses and Syphon Tubes
Drains (Floor Drain Preferable)
Vapor Proof Ventilation
Non-Essential Rooms
Bulking and Impregnating Agent
Storage
Non-Essential Equipment
Bulking Agent Evaporation Tank
Deionized Water
Rain Water Cistern
Wet Artifact Storage Tanks
Bulking or Impregnation Treatment Tank
Dehumidifier and Distilled water catchment
Electrolytic Reduction Tanks with Vapor Proof Exhaust Fans
Sink and Work Bench
Dry Lab Facilities - Artifacts Finished
Humidity Chamber
Chemical Storage
Acids
Bases
Non-Essential Rooms
Office
Dark Room
Bathroom
Artifact Storage
Wood Shop
Classroom
Library
Micoscropy
Non-Essential Equipment
X-Ray
Glass Storage
Microscope
Textile Table
Agitation Tank
Microscope
Freeze Dryer
Stove with Fume Hood
3X Magnification Work Table & Balance Beam Scale
Freezer Refrigerator
Hot Box
Work Bench

Figure 1.

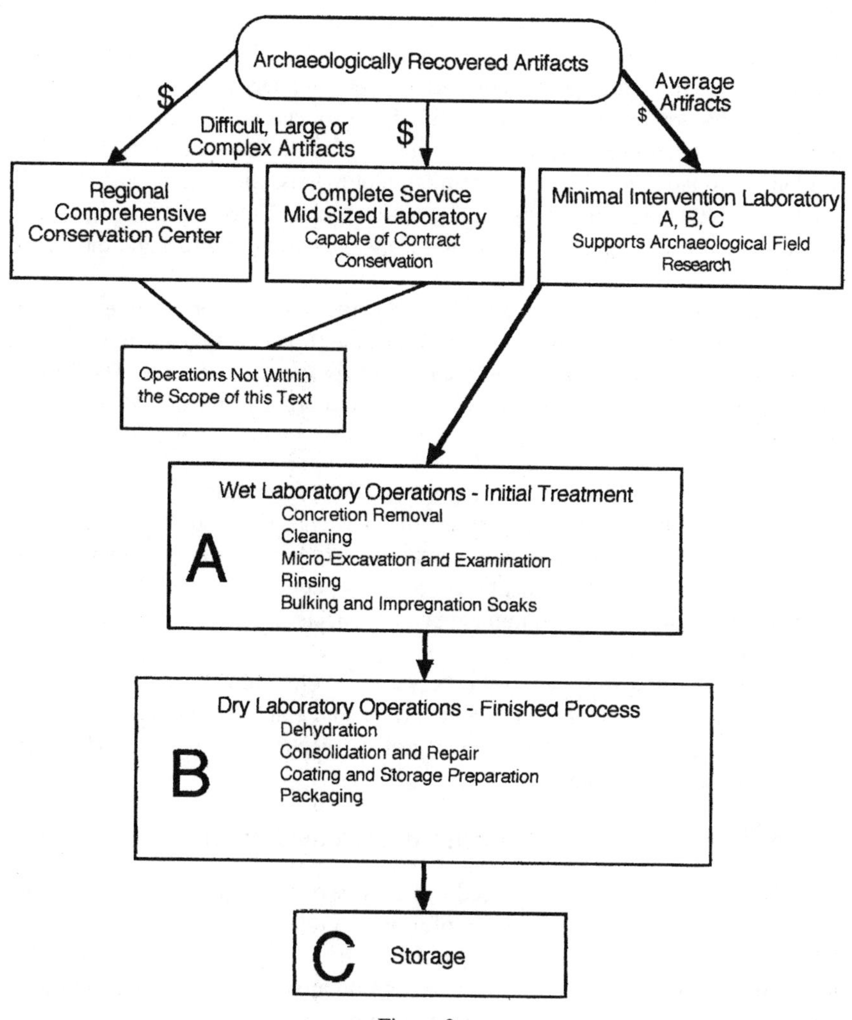
Laboratory Design Philosophy
Archaeologically Recovered Artifacts
$
Difficult, Large or Complex Artifacts
$
Average Artifacts
$
Regional Comprehensive Conservation Center
Complete Service Mid Sized Laboratory
Capable of Contract Conservation
Minimal Intervention Laboratory A, B, C
Supports Archaeological Field Research
Operations Not Within the Scope of this Text
Wet Laboratory Operations - Initial Treatment
A
Concretion Removal
Cleaning
Micro-Excavation and Examination
Rinsing
Bulking and Impregnation Soaks
Dry Laboratory Operations - Finished Process
B
Dehydration
Consolidation and Repair
Coating and Storage Preparation
Packaging
C
Storage

Figure 2.

MINIMAL INTERVENTION LABORATORY (MIL) MISSION

MILs include most academic anthropological laboratories, non-profit archaeological support laboratories, and laboratories set up to support contract archaeology. These aren't necessarily small laboratories in size, but most of their space is devoted to storage of recovered artifacts and very little space is actually designated for actively treating and stabilizing artifacts. A MIL can contain acres of boxes and shelves yet may not have so much as a fume hood or vapor proof ventilation fan to facilitate electrolytic reduction of metals. Cleaned, tagged, bagged, and boxed, is the traditional conservation method for artifacts curated in these labs, a method that needs update and revision if the collections are to stand the test of time.

The methodology outlined in the following chapters for carrying out laboratory objectives are based on a set of ethical and practical codes derived from published conservation sources and accumulated practical experience. The following profile of these principles is set in order of importance and should help shape the objectives of a minimal intervention laboratory (Figure 1, page 10).

1. Health should never be compromised for conservation.
2. All treatments should be reversible (Pearson, 1987; ed. Preface).
3. Use the simplest procedures with the least intervention (Pearson, 1987: 123). Ramp up treatment procedures as necessary.
4. Use multipurpose equipment and chemicals.
5. Do not delay treatment.
6. Work to the ability of the conservator and the resources of the laboratory.
7. All work should be documented (Pearson, 1987; ed. Preface).
8. Normal artifact ageing should not be disguised or removed (Pearson, 1987; ed. Preface).

1. Health Should Never Be Compromised for Conservation

If two or more treatment procedures are shown to be equivalent, or nearly so, the procedure demonstrably less harmful on a personal and environmental level will be chosen in every case. No object or artifact is worth compromising the health or well being of an individual or community. Most treatments and tests today can be conducted in a safe non-polluting manner with the artifact left in a safe condition to handle on completion of treatment.

Reasons for this corollary are easily seen. The minimal intervention laboratory is based on simple procedures and equipment and is not staffed or equipped to deal with storage and use of major amounts of toxic or dangerous substances particularly heavy metal salts used in chemical testing and organic solvents used to dissolve bulking or impregnating agents. Nor is the minimal intervention laboratory ethically able to support expensive and specialized equipment when simple inexpensive equipment has proven, in most cases, to be as effective.

Obviously this corollary is a matter of professional philosophy, and there are times even in the minimal intervention laboratory when flammable solvents and dehydrating agents are used. In these instances the dangers will be explained, and the proper precautionary measures spelled out.

2. All Treatments Should Be Reversible

This is a universally accepted notion in artifact conservation. Viewed in the context of generations, any conservation treatment used today has to be viewed as a temporary measure. Advances in conservation science will dictate new and better treatments in the future and present day approaches will become obsolete. It makes sense, therefore, since any treatment attempted today may have to be reversed, that this eventuality be planned for in advance.

It can be argued that on a molecular level that no treatment is fully reversible. This of course may be true, it is virtually impossible to rinse every molecule of a bulking or impregnating substance from wood cell walls, particularly if the chemical has bonded to the wood fibrils and it may also be impossible to remove every molecule of a coating from a metallic artifact.

Yet, reversibility, as a principle, should be adhered to. The argument against reversibility on grounds that all chemicals cannot all be rinsed is specious. If it is necessary to remove a reversible substance, the few remaining molecules will not interfere with the introduction of new treatment substances. And non-reversible procedures, even those that show promise in the short run, cannot be tested beyond our mortal frame of reference in viewing and caring for artifacts. *The artifacts, in the end, must be able to survive longer than conservation agents placed in, or on, them.*

3. Use the Simplest Procedures with the Least Intervention

The object of artifact conservation treatment is to preserve the appearance, internal microscopic integrity, and dimensions of an artifact as far as possible. Often, at least at a large comprehensive conservation laboratory, there are several conservation methods available per material type.

This is not the case at a minimal intervention lab. A minimal intervention lab will use the easiest, non-toxic procedures and techniques to insure that the laboratory will be cost effective. Minimal intervention labs will not have redundant equipment, multiple chemicals that have similar uses, or technicians trained in complicated, toxic, or costly and superfluous procedures. This will save money, time, effort, and laboratory space. It will also save intangibly by eliminating storage space for chemicals and used toxic chemicals. But even simpler laboratories can ramp up their treatments to satisfy the needs of individual artifacts, either by adding more treatment time, or gradually increasing the strength of solvents and stain removers.

Minimal intervention implies that an artifact will undergo the least cleaning, handling, and treatment needed for its complete stabilization. This will save time and artifact handling per unit making the lab more efficient in artifacts preserved per conservation man/hour. Efficiency can be important in a lab suddenly inundated with artifacts, as is periodically the case.

The simplest procedures do not, by any stretch of the imagination, suggest that the artifact conservation has been compromised. To the contrary, the basis of this manual is the concept that all treatments have been studied according to stabilizing performance, cost, need for equipment, and simplicity. The best treatments, given the parameters of a minimal intervention non-toxic lab, are presented in this manual.

4. Use Multipurpose Equipment and Chemicals

The minimal intervention lab should be set up as efficiently as possible. Laboratory organization suggested in this manual takes into account the varied procedures for different materials and invariably promotes the technique that is most adaptable to different materials. For example, multiple electrolytes (acids and bases) have been suggested in conservation literature depending on type of metal to undergo treatment. Yet arguably, non-toxic sodium carbonate acts equally well for all metallic electrolysis. Sodium carbonate can also be used for metallic artifact pre-treatment, and combined with sodium bicarbonate acts as a rinse and post treatment for several metals. It is an easy choice for electrolyte at the minimal intervention laboratory, saving space, clutter, and complexity.

Polyethylene glycol (PEG) and sucrose are proven non-toxic bulking agents. They do not wear out by the time a treatment is finished and in fact, it has been observed that they remain viable for years after their first use. An efficient laboratory will simply evaporate the water from the treatment solution and use both the PEG and sucrose over again and again. This efficiency will lead to greater laboratory cost savings, simpler techniques, and less clutter for additional work place safety.

5. Do Not Delay Treatment

Conservation laboratories cannot always keep up with the volume of material excavated. This is particularly true when conservation planning has not been thoroughly calculated into the research design of a project. Backlogs of material and delayed treatment times become the unfortunate outcome of such planning. Treatments may also be delayed because some individual artifacts pose incredibly difficult conservation dilemmas and are placed on the "back burner" while less difficult to treat artifacts are finished.

As mentioned, most artifacts reach an electro-chemical equilibrium with their environment over time. Excavation and or removal of an artifact from this balanced environment upsets this equilibrium and usually begins the decay process at an accelerated pace. Artifacts often survive for centuries on site better than a few months or years in storage awaiting treatment. It is the conservator's duty to process artifacts in a timely manner. If treatment delay is a reoccurring problem, overall project research design and site excavation methodology should be reevaluated.

6. Work to the Ability of the Conservator and the Resources of the Laboratory

This ethical code is a real test of personality. Some conservation technicians will take on projects that are well beyond their experience and expertise, while others will call for help immediately when the unexpected happens. Minimal intervention lab technicians need to work between these extremes. Working beyond experience level can lead to disaster, and cannot be recommended for valuable, important, or non-robust artifacts. On the other hand, valuable experience can be acquired through technical lab practice in preserving multiple redundant artifacts (example: a keg of nails). Chances are, even practice or classroom training stabilization is more attention than most artifacts will ever receive.

All laboratories, however, have their limits. A laboratory fitted with the finest equipment is only as good as its conservator and no conservator has done and seen everything. The best conservator's know when an artifact is beyond their experience or the accumulated experience of their research library and will network with other conservators to find a solution, or refer an artifact to a more specialized laboratory.

7. All Work Should Be Documented

In the long view of things even the best conservation methods are probably no more than temporary fixes. Even the most stable artifact may undergo changes in storage. Few conservators would be foolish enough to guarantee their treatments will continue to stabilize an artifact forever; chemical agents break down over time with minute variations in humidity, light intensity or handling. This is another strong argument for corollary number two, that all conservation techniques should be reversible. The first step in re-treating an artifact would be the removal of whatever remains of the original stabilizing or coating material.

Conservation documentation, or the idea of recording all of the steps used to store, clean, and treat an artifact is a vital step in its life long curation. This will allow conservation technicians to plan further conservation treatment and reverse any adverse affects of the original treatment attempt.

8. Normal Artifact Aging Should Not Be Disguised or Removed

As far as possible original dyes, paints, inks, stains, gilding, wear marks, tool marks, the macro, and micro-structure of an artifact, should not be changed or removed. Decayed parts of an artifact should be stabilized and conserved, not replaced. If it is necessary to replace a section of an artifact for mechanical strength or stabilization, it should be done of material that can easily be removed and contrasts in color or texture significantly from the original artifact in order that observers not mistakenly conclude the artifact is intact.

These principles guide the choice of conservation technique and procedure at a minimal intervention laboratory and should allow conservators to treat artifacts against decay while also protecting themselves and the environment. These codes also act as a filter from which the hundreds of treatments available can be reduced to only those few that meet the needs and obligations of the minimal intervention laboratory. If another principle is needed to supplement the first eight, it would be that conservation is 50% informed inventiveness and 50% common sense.

WET LAB—ARTIFACTS ENTER

Minimal intervention artifact conservation facilities can be fabricated virtually anywhere or of any size building or room so long as there is adequate floor space to meet the design needs of the lab. Regardless of the size of the laboratory it is usually necessary to divide whatever space is available into Dry Lab Facilities and Wet Lab Facilities in order to separate the two major environmental components of conservation. Most household kitchens would make adequate labs and already contain most of the tools and equipment necessary for minimal intervention conservation. The hypothetical laboratory shown in Figure 1 could be described as no more than a kitchen (Dry Lab Facility) and a garage or storage area (Wet Lab Facility).

Generally speaking, artifacts are brought from the field into the Wet lab facility for storage, cleaning, and treatment, after which they pass to the dry lab facility for dehydration and protective coating. Most conservation treatments use three stages depending on material type. The first is a deep cleaning or sometimes an electro-chemical cleaning of an artifact. This removes contaminants from the interior as well as the surface of an object. Neutral chemical agents are at times added to maintain an object's shape. Finally applications of consolidants and protectants strengthen the artifact and guard against further oxidation or hydration. Most of the deep cleaning and the introduction of bulking and impregnating agents to maintain an artifact's shape are procedures done in the wet lab.

An examination of the wet lab facilities in Figures 1 & 2 (pages 10 and 11) reveals that most of the area is occupied by artifact pre-treatment storage. Metallic artifacts and those recovered from wet environments need to be kept wet in a stabilizing solution while non-metallic artifacts recovered from dry environments need to be kept dry until treatment commences. Laboratories that will treat both wet and dry recovered material will need both tanks and dry shelf storage.

Nearly any type of tank will do for wet artifact storage including plastic or galvanized metal. Agricultural feed troughs can be large and cost effective. It should be noted that if artifacts are stored in metal tanks that the tank should be lined in plastic or the artifacts placed in non-conductive containers to prevent galvanic coupling (see Chapters 3, 4, and 5). Galvanic coupling occurs when two dissimilar metals come into contact to create corrosion via electron transfer. Metal tanks are also easier to maintain if they are raised above the floor on wooden, cinder block, or brick platforms to prevent corrosion induced by floor spills and cleaning.

Shelving should be strong, stable and fastened to a wall or floor to prevent accidental destruction of stored artifacts. Though any material will do for shelving it should be noted that use of metal shelving will require that metal artifacts be placed on non-conducting platforms of wood or plastic. Artifacts should not touch other artifacts while in pre-treatment storage and should generally be organized by material type.

The second largest space used in the wet lab is for sinks and workbenches. This also includes space set aside outdoors for concretion removal and rough artifact cleaning, a particularly dirty task best done well away from the rest of the facility. The compressor is used to operate air scribes and bead blasters, two very good methods for mechanically cleaning metals. If floor space is at a premium, vertical air compressors take up little space. Running water is essential in the wet lab to fill storage and treatment tanks. Running water is also essential in many of the rinsing processes that take place in the wet lab.

Water is one commodity the wet lab cannot have too much of. The wet lab facilities on page 10, have four types of water, all of which are useful in various treatments. Tap water can be used in initial rinses and for cleaning the facility (a never-ending task in itself). But tap water contains varying amounts of chlorine (depending on the municipality). Chlorine or chloride ions, even in small amounts, are particularly destructive in the breakdown of certain materials including metals, ceramics, and organics. Rinsing chlorides from artifacts is a large part of many treatment processes. Therefore, both distilled and deionized water are manufactured in the wet lab and stored in containers for ready use.

Distilled and deionized water are virtually interchangeable in many rinse processes since they both contain very few chloride ions. Both can be manufactured with elaborate scientific equipment or by fairly simple arrangements. Deionized water is produced by filtering tap water through a deionizing unit that consists of a plastic cartridge in which tap water is introduced at one end and deionized water emerges from the other. These units are cheap, simple to operate, and are available through scientific equipment catalogues. Stills to produce distilled water are also available through scientific equipment catalogues but tend to be expensive. An alternative is to collect the water produced by a dehumidifier or air conditioning unit. This water, depending on the cleanliness of the coils in the unit, is virtually chloride free and will suit the purposes of minimal intervention lab for a fraction of the cost of distilling water.

Another low chloride water source is rainwater. Rainwater tends to contain some chlorides (in coastal areas), but not nearly the amount found in most tap water. Rainwater is free, and acts as a good intermediate phase rinse solution between tap water and deionized and distilled water. Cisterns to catch rainwater are easily constructed of a plastic barrel situated at the bottom of a roof's gutter and rainspout.

Solvent storage for acetone and denatured or methyl alcohol will also be needed in either the wet or dry lab facilities. Solvent storage is subject to building and chemical codes and may need to be stored in a flame resistant metal locker designated to hold flammable liquids. It can also be contained in portable flammable liquid containers with the contents clearly marked on both the cabinet and the container.

Additional needs of the wet lab include a vapor proof ventilation system that pulls fresh air into the lab and exhausts the internal air outside. The ventilation system does not have to be powerful, just strong enough to circulate the laboratory air on a regular basis. Minute amounts of hydrogen and oxygen are given off during the electrolytic rinsing process for metals. Though the amounts of these gases given off are negligible their concentration in one area could prove explosive should there be a trigger. Smoking and production of flame of any kind should be discouraged in the wet lab.

The hypothetical facilities on page 7 contain some equipment that a minimal intervention facility may not need, but are deemed extremely useful. For example, a MIL can do without a compressor with bead blasting and air scribe capabilities, this type of work can be done with time and effort using hand tools (dental picks, scalpels, or an electric scribe), it is simply much more convenient and cost effective to operate the compressor and its accessories if they are available. Labs often start as bare-bones facilities and upgrade when funding becomes available. A lift for heavy artifacts and a means to carry heavy bags of electrolyte or barrels of bulking agent will also be convenient. Hydraulic engine hoists for heavy artifacts and hand trucks are economical and more than sufficient to move supplies. Those unfamiliar with artifact conservation are often surprised to find that some areas of an artifact conservation facility are far removed from the stereotypical laboratory and create a setting more akin to a factory. In this environment common sense concerning the movement and storage of both artifacts and conservation supplies is a must. Common sense must also apply to layout and organization of the wet lab both for the efficiency of the lab and the safety of the conservation technicians working there. Storage of supplies is not shown in the hypothetical wet lab diagram, and of course it is completely dependent on the facility but much can be done in any lab to cut down inefficiency by removing redundant chemicals and recycling others.

Chemicals and bulking agents such as polyethylene glycol and sucrose can be reused time and again. The Bulking Agent Recovery Tank (BART) and storage system shown below will recycle PEG and sucrose multiple times. BART is a large system but it demonstrates how a similar yet smaller system could be used at a minimal intervention lab to recover and recycle valuable bulking agent, while at the same time limiting the amount of storage space needed for chemical agents.

Finally the Bulking or Impregnation Treatment Tank can be fabricated from a metal agricultural tank wrapped outside in foil bubble wrap insulation and

Figure 3. The Bulking Agent Recovery Tank (BART) system at the artifact laboratory, Program in Maritime Studies, East Carolina University. Bulking and impregnating agents are filtered, pumped into the central tank where the water is evaporated through a central vent fan. The agents are then gravity feed to storage tanks on either side of the evaporation tank. PEG and sucrose can be used for many years. Photograph by Chris Valvano.

supported off the floor on wooden blocks or pallets. Wooden artifacts will soak in bulking or impregnating agents for up to a year duration as they absorb a specific amount of PEG or sucrose. Though the tank will work simply by placing the artifacts and chemical in it, it will work faster and more efficiently if it is heated, circulated, insulated, and covered. Scientific heaters and circulators are available for this purpose at tremendous cost. Alternative heaters and stirring devices are easily obtained through the imaginative use of restaurant sink heaters, waterbed heaters, and fish tank filters. Other very usable stirring devices include electric fish trolling motors; inexpensive to buy, reliable for years, and easily powered by a10 amp battery charger. For safety sake the leads to the trolling motor should be spliced permanently into the battery charger cables; positive to positive (red) and negative to negative (black).

When artifacts have been cleaned and treated in the wet lab it is time for their transfer to the dry lab for final treatment and dehydration. The two-lab concept is simply that:—a concept. In reality the two labs can be in the same room or the same building, it is the atmospheric separation of processes that is most important and easily brought about by physical separation.

DRY LAB FACILITIES—ARTIFACTS FINISHED

The dry lab facilities depicted in Figure 1 are required to dehydrate, consolidate, and place a protective coating on objects that have been deep cleaned and treated in the wet lab. The dry lab is also the area where an artifact can be closely examined for maker's marks, fabrication detail, wear patterns, and other micro-excavation details. A magnification table complete with 3X lighted magnification lens and microscope is a necessity for careful examination of artifacts but not completely necessary for the conservation and stabilization of an artifact. A balance beam is also a convenient tool to weigh artifacts during analysis and is perhaps more a conservation necessity than either of the aforementioned items. Yet all of these instruments are very useful and by necessity have to remain separate from the damp and sometimes dusty conditions that prevail in the wet lab. Should space and equipment be available some additional but non-essential facilities could include offices, a dark room, bathroom, artifact storage, a wood shop, a classroom and library, and separate microscope lab.

The heart of the dry lab facilities is the enigmatic stove and oven that combine with the freezer refrigerator to give the facility its kitchen like atmosphere. In academic settings the stove and oven are being replaced with scientific versions that serve the same function at many times the expense. The range hood is also being replaced with a scientific fume hood. The oven will be used to dehydrate metals and desiccate other material while the burners will be used to melt microcrystalline wax used to dehydrate and coat metals.

The refrigerator acts as a store for recovered organic material while the freezer will be used to freeze-dry selected organic and other pre-treated artifacts. The vaunted vacuum freeze drying systems, seen in the 1980s as magic conservation bullets, proved expensive and only as successful as their pre-treatment, certainly little better in practical terms than atmospheric freeze-drying. In theory freeze-drying allows frozen water in an object to evaporate gradually through sublimation, turning directly from a solid to a gas. Sublimation bypasses the liquid phase that creates so many problems for artifacts (see Chapter 2 and 7 waterlogged Wood and Organics Other than Wood). Vacuum freeze-drying never lived up to its potential and its initial costs are prohibitive for a minimal intervention laboratory. Less expensive equipment such as the Relative Humidity Chamber (RHC) has demonstrated very good results in the dehydration of pre-treated wood and organic material. The relative humidity chamber can be built in house, at a fraction of the cost of a vacuum freeze-dryer.

The RHC dehydrates pre-treated organic material and wood by allowing it to slowly dry. Slow drying after a pre-treatment in bulking or impregnating agents lessen the effects of sudden drying (see Chapter 2 Waterlogged Wood) and permits a gradual reintroduction of the artifact to a normal atmosphere. A relative humidity chamber can be any sort of large container, preferably plastic, that can be opened to receive an artifact and then sealed to contain humidity. A plastic agricultural tank cut out and fitted with a shower curtain over the door works well. Ideally the chamber would contain a humidifier that can be turned on from outside the chamber (operation of the chamber is explained in Chapter 2).

The Hot-Box is another specialized piece of equipment that is easily fabricated from an old refrigerator or freezer that no longer works. A light bulb is installed in the hot box that acts as a heat source. Temperature control is a matter of adjusting the light bulb wattage, the higher the wattage the hotter the box. The Hot-Box will work well for bulking smaller wooden or organic objects that don't need the large volume treatment tank located in the wet lab.

Chemical storage is essential in the dry lab. Acids or oxidizing agents are located on separate shelving from reducing agents (bases) to lessen the chance that a spill will result in disaster. Storage shelves for these chemicals are also strong and secured to the wall or floor. All chemical containers on the shelves of these storage units should be clearly marked for content and rest in chemically resistant spill trays in case of leakage.

SAFETY

Safety in the conservation laboratory is largely a matter of common sense, organization, and tidiness. Safety issues at the minimal intervention laboratory are not as great as those in more sophisticated laboratories. But safety does remain an issue even at a minimal intervention lab with few toxic substances. This is particularly true since falls and eye injuries remain the largest type of accident reported in a lab. An uncoiled hose, or a beaker filled with a solvent placed on the edge of a counter top can always be a problem given the proper unlucky circumstances. Fortunately, proper safety habits can cut down on the risks and accidents can usually be avoided before they happen with proper common sense habits, like picking the place up after work has been done.

The first safety rule in any lab is the wearing of safety glasses. Safety glasses not only protect the wearer from injury from direct flying objects, but perhaps most importantly, from objects that are flying from unknown angles. This is particularly important in a small lab where the actions of one technician flaking concretion can easily injure another technician nearby. Eye wash stations are also a must in both the dry and wet lab. These can be inexpensive portable units that hang in clearly marked wall spaces or they can be built into the plumbing of the laboratories themselves.

Rubber gloves, protective coveralls, and breathing masks are also a good idea depending on the processes and chemicals that are being handled. Even a low toxicity lab will contain some chemicals or solvents that can irritate eyes, skin or throat. In the end it is a good habit to treat all chemicals and substances in the lab with respect and care. Sloppy technicians will injure unintentionally by spilling substances and not cleaning after themselves. The next technician will unwittingly rub their eye after encountering a days old spill. Cleanliness is the second safety rule at the minimal intervention lab.

Every laboratory should maintain an inventory of every chemical on site. Data sheets for these chemicals can be obtained from the health and safety unit assigned to your lab and Material Safety Data Sheets (MSDS) will be issued for each of these chemicals. The MSDS will list the toxicity and characteristics of these chemicals and should be consulted should any questions arise concerning a substance that will be used or should a technician attempt a new procedure. Posting and use of the MSDS is rule three.

Fire extinguishers are placed at strategic points in both labs and near each clearly marked exit. Properly trained safety technicians will check fire extinguishers on a regular basis and their inspection dates posted on the fire extinguisher. Melted wax treatments on the stove will be under continual surveillance from the technician performing the procedure and the stovetop can NEVER be left on while unattended.

Where possible in this manual safety issues will be brought up in the general discussion of treatment procedure but no manual or warning can take the place of common sense in both the arrangement and operation of a laboratory. Safety is also an ongoing issue that is enhanced with the continual feed back of the technicians working in a lab. It does no good to observe a problem without suggesting a solution and acting on it. It is a cliché, but safety in a laboratory is the responsibility of every person that walks in the door.

CREATING THE MINIMAL INTERVENTION LABORATORY

The suggestions outlined in this chapter are intended as a general guide, not as a technical plan to create a laboratory. Imagination, know-how, and common sense will permit conservation technicians to create or add laboratory space and capability as they see fit or as money and time permit. The theoretical minimal intervention laboratory diagramed in Figure 1 and described in this chapter is simply a pattern or model, just as *The Archaeologist Manual for Conservation* is a tool. A laboratory is always more a process than a structure. It grows and changes as the conservators and technicians gain experience and knowledge, or as is often the case; when they have to grow to support the expanding needs of an archaeological program.

It is hoped this manual will give conservation technicians and conservators a new instrument from which they can expand their concept of conservation and implement new conservation techniques or refine those already in place. Vast numbers of archaeological artifacts lie in storeroom boxes, seemingly out of sight and out of mind, but few have truly been stabilized through surface cleaning and storage. This manual should permit archaeological conservators to reevaluate conservation at their laboratories add new techniques and methodology to their repertoire of procedures, or create a facility capable of conserving artifacts recovered from virtually any site environment.

BIBLIOGRAPHY

The minimal intervention laboratory concept is simplistic but not easily implemented. Its underpinnings begin with comprehensive accumulated research, represented in this manual in the form of bibliographies located at the end of each chapter. This chapter's bibliography contains general archaeological conservation sources including all of the new and classic conservation texts, plus accumulated articles and technical notes gleaned from technical journals and periodicals. Succeeding chapter bibliographies contain books and articles concerned only with the type of material covered in that chapter.

CHAPTER 1: THE MINIMAL INTERVENTION LABORATORY

Abinion, Orlando V. "The Recovery of the 12th century Wooden Boats in the Philippines." *Bulletin of the Australian Institute for Maritime Archaeology* 13 (1989): 1–2.

Allison, Rutherford. "Infrared Photography Using an Ultraviolet Radiation Source." *Medical and Biological Illustration*. Vol. 27, 1977: 35–36. [ECU-521]

Anon. *Archaeological Conservation Newsletter*. (South Carolina State Museum Conservation Laboratory Publication.) 2 (January 1990). [ECU-344]

Anon. "Conservation in Archaeology and the Applied Arts." *ICOM Congress*. Stockholm, (1975).

Anon. "Fabric Protects Historic Ship." *Textile Month* (1983): 23. [ECU-217]

Anon. "On Site Conservation Requirements for Marine Archaeological Excavations." *International Journal of Nautical Archaeology* 6 (1977): 37–46.

Anon. "Underwater Treasures Under the Microscope: Objects from the *DeBrack* Analyzed at Winterthur." *Winterthur Newsletter* 34 (n.d.): 14–15.

Applebaum, Barbara. "Comments from a Conservator on Climate Control." *WAAC Newsletter* Vol. 18 No. 3, 1996. [ECU-749]

Arthur, B.V. "The Conservation of Artifacts from the Wreck *Machault*." *ICOM Proceedings* (3rd Triennial Meeting, Madrid) (1971).

Ashley-Smith, Jonathan. *Risk Assessment for Object Conservation*. London: Elsevier, 1999.

Barclay, Robert. "The History of Conservation: The Contribution of Michael Faraday." *CCI Newsletter* (Canadian Conservation Institute, Ottawa) No. 29, 2002. [ECU-612]

—. "Mrs. Beeton, Household Conservator." *CCI Newsletter* (Canadian Conservation Institute, Ottawa) No. 28 2001. [ECU-613]

—. "Staying in Tune with Conservation." *CCI Newsletter* (Canadian Conservation Institute, Ottawa) No. 22 1998. [ECU-614]

Barkman, L. "Conservation of Finds." *Proceedings* (First Southern Hemisphere Conference on Maritime Archaeology) (1978): 119–120.

—. "Inventory of Ancient Monuments Underwater in Sweden." *Proceedings* (First Southern Hemisphere Conference on Maritime Archaeology) (1978).

—. "Treatment of Waterlogged Finds." *Proceedings* (First Southern Hemisphere Conference on Maritime Archaeology) (1978): 120–126.

—. "The Conservation of the Waterlogged Warship *Wasa*." *Proceedings* (ICOMOS Symposium on the Weathering of Wood, Ludwigsburg.) (1969): 99–106.

—. "The Preservation of the Warship *Wasa*." *Maritime Monographs and Reports* (National Maritime Museum, Greenwich, U.K.) 16 (1975), 65–105. [ECU-251]

—. "Treatment of Waterlogged Finds." *Proceedings* (First Southern Hemisphere Conference on Maritime Archaeology) (1977): 120–126.

—. *The Preservation of the Wasa*. Stockholm Museum, 1967.
Barkman, L. and A. Frantzen. "The *Wasa*: Preservation and Conservation." *Underwater Archaeology, A Nascent Discipline* (UNESCO) (1972): 231–242.
Bass, George F. "The Shipwreck at Serce Limon, Turkey." *Archaeology* 32 (1979): 36–44. [ECU-0270]
Bengtsson, Sven. "What about *Wasa*?" *Realia* 1 (1990): 309–311.
Berrett, Kory. "Conservation Surveys" Ethical Issues and Standards." *Journal of the American Institute for Conservation*. Vol. 33 No. 2, 1994, pp. 193–198. [ECU615]
Black, J., ed. *Recent Advances in the Conservation and Analysis of Artifacts*. (Jubilee Conservation Conference Papers. University of London, Institute of Archaeology, Summer Schools Press, London) (1987).
—. "The Biodeterioration and Conservation of Archaeological Materials Symposium." (Abstracts, University of Portsmouth, Portsmouth) (September 1992). [ECU-435]
Borden, C.E. *Excavation of Water Saturated Archaeological Sites*. Natural Museums of Canada 1976.
Borgin, K. "Mombasa Wreck Excavation: Second Preliminary Report, 1978. Appendix 2: Progress Report on Evaluating the Thessaloniki Process." *International Journal of Nautical Archaeology* (1978): 301–319.
Bourque, B.J., S.W. Brooke, R. Kley, and K. Morris. "Conservation in Archaeology: Moving Toward Closer Cooperation." *American Antiquity* 45:4 (1980): 794–799.
Bried, Raymond. "She Wasn't Made to be Taken Apart." *Yankee* (1973): 84–93, 106–111.
Bright, Leslie S., W.H. Rowland, and I.C. Bardon. "Preservation of the *Neuse*, Appendix l." *A Question of Iron and Time* (North Carolina Division of Archives and History) (1981). [ECU-16]
Brooke, S. and K. Morris. "A Preliminary Report on the Conservation of Marine Archaeological Objects from the *Defense*." (Bulletin of the American Institute for Conservation, Preprints, Boston) (1977): 27–34.
Buck, Richard. "On Conservation." *Museum News* 52 (1973): 15–16. [ECU-327]
Bump, Herbert D. "Report Concerning the Condition of Artifacts at the Confederate Naval Museum, Columbus, Georgia." (Research amd Conservation Laboratory Monograph, Florida Department of State) [n.d.] [ECU-0376]
Byung-Mo, K., K. Ki-Woong, and P.H. Kieth. "The Shinan Shipwreck." *Museum* 35 (1983): 35–37.
Canadian Conservation Institute. *Annual Report 1995–1996*. Canadian (1996). [ECU-527]
Canadian Conservation Institute. *Annual Report 1994–1995*. Canadian (1995). [ECU-554]
Canadian Conservation Institute. *Annual Report 1993–1994*. Canadian (1994). [ECU-519]
Canadian Conservation Institute. *Catalog of Publications*. 1995. [ECU-555]
Canadian Conservation Institute. *CCI Newsletter* September 1996, No. 18. [ECU-528]
Canadian Conservation Institute. *CCI Newsletter* (1987–present). [ECU-298]
Canadian Conservation Institute. *CCI Newsletter* March 1997, No. 19. [ECU-586]
CCI Laboratory Staff. "Encapsulation." *CCI Notes* (Canadian Conservation Institute) 11/10 (1995). [ECU-543]
CCI Laboratory Staff. "Emergency Preparedness for Cultural Institutions: Introduction." *CCI Notes* (Canadian Conservation Institute) 41/1 (1995). [ECU-547]
CCI Laboratory Staff. "Emergency Preparedness for Cultural Institutions: Identifying and Reducing Hazards" *CCI Notes* (Canadian Conservation Institute) 14/2 (1995). [ECU-548]
CCI Laboratory Staff. "Planning for Disaster Management: Introduction." *CCI Notes* (Canadian Conservation Institute) 14/1 (1988). [ECU-443]
CCI Laboratory Staff. "Precautions for Storage Areas." *CCI Notes* (Canadian Conservation Institute) 1/1 (1992). [ECU-458]
CCI Laboratory Staff. "Time Capsules." *CCI Notes* (Canadian Conservation Institute) 1/6 (1995). [ECU-535]
CCI Laboratory Staff. "Track Lighting." *CCI Notes* (Canadian Conservation Institute) 2/3 (1992). [ECU-459]
CCI Laboratory Staff. "Using a Camera to Measure Light Levels." *CCI Notes* (Canadian Conservation Institute) 2/5 (1992). [ECU-460]

CCI Laboratory Staff. "CCI Environmental Monitoring Equipment." *CCI Notes* (Canadian Conservation Institute) 2/4 (1993). [ECU-471]

CCI Laboratory Staff. "Ultraviolet Filters." *CCI Notes* (Canadian Conservation Institute) 2/1 (1994). [ECU-472]

CCI Laboratory Staff. "The Diphenylamine Spot Test for Cellulose Nitrate in Museum Objects." *CCI Notes* (Canadian Conservation Institute 17/2) (1994). [ECU-509]

Carrell, Toni L., ed. *Underwater Archaeology Proceedings* (Society of Historical Archaeology Conference, Tucson, Arizona) (1990).

"Conservation in Archaeology and the Applied Arts." *ICOM Congress* (Stockholm) (1975).

"Conservation Center Stage" *Museum News* Vol. 76, 1997, pp. 48–49, 59–63. [ECU-750]

"Conservation in Nautical Archaeology." *INA Newsletter* Vol. 6 No. 1, Spring 1979. [ECU-579]

Cornet, I. *Scientific Methods in Medieval Archaeolgy*. Sacramento: University of California Press, 1970.

Costain, Charles. "The Light Damage Slide Rule." *CCI Newsletter* (February, 1989): 19. [ECU-24]

Cowan, Rex, Z. and Peter Marsden. "The Dutch East-Indiaman *Hollandia* Wrecked on the Isles of Scilly in 1743." *International Journal of Nautical Archaeology* 4 (1975): 267–300. [ECU-380]

Cronyn, J.M. *The Elements of Archaeological Conservation*. London: Routledge, 1990. [ECU-350]

Daley, Thomas W. and Lorne D. Murdock. "Excavating and Raising Artifacts from a Marine Environment." [ECU-556]

—. "Underwater Molding of a Cross Section of the San Juan Hull—Red Bay, Labrador." *ICOM 7th Triennial Meeting*, Sept. 1984. [ECU-526]

Daugherty, R. "Responsibilities and Commitment in the Excavation of Wet Sites." *Proceedings* (International Conference on Wet Site Archaeology, University of Florida, Gainesville) (1986).

Davies, P.N. "The Discovery and Excavation of the Royal Yacht Navy." *Maritime Wales* (1978): 25–32.

Davison, S. "Antiquities." *Adhesives and Consolidants, Reprints of the Contributions to the Paris Congress 2–8 Sept. 1984*. IIC London, 1994. [ECU-563]

Dawson, John E. and Thomas J.K. Strang. "Controlling Vertebrate Pests in Museums." *CCI Technical Bulletin* (Canadian Conservation Institute, Ottawa) 13 (1991). [ECU-375]

Dawson, John E. and Thomas J.K. Strang. "Solving Museum Insect Problems: Chemical Control." *CCI Technical Bulletin* (Canada Conservation Institute, Ottawa) 15 (1992). [ECU-434]

"Dechlorination in Buffered Citrate." [ECU-587]

Degrigny, Christian. *Bulletin of the Research on Metal Conservation*. ICOM Metal Working Group, February 2003. [ECU-608]

"Description and Evaluation of Florida State Preservation Laboratory." [ECU-588]

Denton, Mark H. and Joan S. Gardner. "Recovery and Conservation of Waterlogged Goods from the Well Excavated at the Fort Loudon Site, Fort Loudon, Pennylvania." *Historical Archaeology* 17:1 (1983): 97–103. [ECU-382]

De Jong, J. "The Conservation of Shipwrecks." *ICOM Proceedings* (Zagreb) (1978).

—. "Protection and Conservation of Shipwrecks." *Greenwich Archaeological Series No. 5* (National Maritime Museum, British Archaeology Reports, International Series 66) (1979): 1–10.

De Jong, J., W. Eenkhoorn, and A.J.M. Wevers. "Controlled Drying as an Approach to the Conservation of Shipwrecks." *ICOM-WWWG Proceedings* (Ottawa) (1981): 1–10.

Doran, Glen H., and David N. Dickel. "Multidisciplinary Investigations at the Windover Site." *Proceedings* (First International Conference on Wet Site Archaeology, University of Florida, Gainesville) (1986).

Doroman, E.A. *Conservation in Field Archaeology*. London: Metheun, 1970.

—. *The Excavation of Water-Saturated Archaeological Sites (Wet-Sites) on the Northwest Coast of North America* (Archaeological Survey of Canada Mercury Series, No. 50, National Museum of Canada) (1976).

Down, Jane L. "CCI Analysis Aids Conservation of the Archimedes Palimpsest." *CCI Newsletter* (Canadian Conservation Institute) No. 28, 2001. [ECU-617]

Erhardt, David and Marion Mecklenburg. "Relative Humidity Re-Examined" *Ottawa Congress of the IIC*, 1994. [ECU-751]

Erhardt, David, Marion F. Mecklenburg, Charles S. Tumosa, Mark McCormick—Goodhart. "The Determination of Allowable RH Fluctuations" *Smithsonian Magazine* Vol. 17 No. 1, 1995. [ECU-752]

The Excavation of Water-Saturated Archaeological Sites (Wet-Sites) on the Northwest Coast of North America, Archaeological Survey of Canada Mercury Series, No. 50, National Museum of Canada (1976).

—. "The Determination of Appropriate Museum Environments." *British Museum Occasional Paper* Num. 16, pp. 153–163. [ECU-753]

Fawcett, Jane ed. *Historic Floors, their Care and Conservation*. London: Elsevier, 2001.

Feilden, Ernard. *Conservation of Historic Buildings*. London: Elsevier, 2003.

Feller, Robert. "Aspects of Chemical Research in Conservation: The Deterioration Process." *Journal of the American Institute for Conservation*. Vol. 33 No. 2, 1994. [ECU-618]

Ferrier, R.J. and P.M. Collins. *Monosaccharide Chemistry*. London: Penguin Books, 1972.

Folan, W.J., J.H. Rick, and W. Zachaschuck. "The Mechanization of Artifact Processing." *American Antiquity* 33 (1968): 86–89.

Florian, M-L. E. "The Underwater Environment." [ECU-590]

"Freeze Drying Services." Hackettstown, NJ: The Artifact Research Center. [ECU-591]

Fruter, E.V. *Introduction to Natural Protein Fibers: Basic Chemistry*. New York: Barnes and Noble, 1973.

Gannon, Robert. "What Really Sank the *Titanic?*" *Popular Science*. Feb. 1995: 49–55, 83. [ECU-522]

Garfield, Donald. "Hot Topic, Cool Reception." *Museum News*, 1995. [ECU-754]

Garcia-Vallés, F. Blazquez, J. Molera, and M. Vendrell-Saz. "Studies of patinas and Decay Mechanisms Leading to the Restoration of Santa Maria de Montblanc (Catalonnia, Spain)" *Studies in Conservation*. Vol. 41 No. 1, 1996. [ECU-619]

Gilliand, Marion S. "Marco's Buried Treasure: Wetlands Archaeology and Adventure in Nineteenth Century Florida." *Wet Site Archaeology* B.A. Purdy, ed., (1988): 255–261.

Gould, Donald. "Draining the Zuider Zee Uncovers a Boneyard of Ancient Ships." *Smithsonian* 5 (1974): 66–73. [ECU-276]

Graham-Bell, Maggie. *Preventative Conservation: A Manual*. Victoria, British Columbia: British Columbia Museums Association, 1983. [ECU-1]

Grattan, David. "New Directions in Education for Marine Archaeological Conservation." *CCI Newsletter* (Canadian Conservation Institute) No. 28, 2001. [ECU-620]

Green, Jeremy N. "Field Conservation in Marine Archaeology: A Consumer's View Point." *Proceedings* (National Seminar on Cultural Material, Perth, Australia) (1976): 195–197.

—. "The VOC Ship *Batavia* Wrecked in 1629 on the Houtman Abrolhos, Western Australia," *International Journal of Nautical Archaeology* 4 (1975): 43–63. [ECU-164]

Green, Lorna R. and Vincent Daniels. "Shades of the Past." *Chemistry in Britain* Vol. 31 No. 8, 1995. [ECU-656]

Greenhill, Basil. "Vessel of the Baltic: The Hansa Cog and the Viking Tradition." *Country Life* (1980): 402–404. [ECU-132]

Groen, C. "The Use of Tomography in the Analysis of Maritime Archaeological Material." *Proceedings* (First Southern Hemisphere Conference on Maritime Archaeology) (1978): 144–145.

Gulbeck, Per E. *The Care of Antiques and Historical Collections*. American Association for State and Local History, 1985. [ECU-3]

Gutierrez, Gilles. *Texinfine Product Brief* (French Ministry of Research and Technology, Lyon, France) [n.d.] [ECU-379]

Hallowall, C. "Disappearance of the Historic Ship *Tijger*," *Natural History* 83 [n.d.]: 12–28. [ECU-158]

Hamilton, Donny L. "Preliminary Report on the Archaeological Investigation of the Submerged Remains of Port-Royal, Jamaica, 1981–1982." *International Journal of Nautical Archaeology* 13 (1984): 11–15.

—. "Conservation of Cultural Resources, I." (Class Handouts for ANTH 605, Texan A&M University, Nautical Archaeology Program, August) (1988). [ECU-364]

—. *Basic Methods of Conserving Underwater Archaeological Material Culture*. Washington, D.C.: U.S. Department of Defense Legacy Resource Management Program, January 1996. [ECU-610]

Harrison, R.F. "The *Mary Rose* Tudor Ship Museum." *Museum* (France) 35 (1983): 44–49.

Hauser, Robert, "Technology Update: Enzymes in Conservation—A Conference Report," *Technology and Conservation*, Vol. 11 No. 4 (Winter 1992–93): 13–15.

Hay, James. "The History of Conservation: Furniture Polishes." *CCI Newsletter* (Canadian Conservation Institute) No. 27, 2001. [ECU-621]

Hon, David N.S. "An Implementation Strategy for Stabilization of the U.S.S. *Cairo*." (Progress Report, No. CA-5000-4-8003, College of Forest and Recreation Resources, Clemson University) [n.d.] [ECU-377]

Horie, C.V. *Materials For Conservation*. London: Elsevier, 1987.

Hough, Mary Piper. "What are Appropriate Standards for the Indoor Envrionment, Review of Symposium." *WAAC Newsletter* Vol. 17 No. 3, 1995. [ECU-755]Hurt, Joseph L. "Preliminary Report: Preservation of the CSS *Muscogee*." (Georgia Historical Commission) (1966): 1–14. [ECU-42]

ICOM Committee for Conservation Working Group on Wet Organic Archaeological Materials Newsletter. Nos. 25, 28, 31, 32. [ECU-593]

Iskander, Z. *The Cheops Boat, Part 1*. Cairo: Antiquities Department of Egypt, 1960.

Jenssen, V. and Colin Pearson. "Enviromental Considerations for Storage and Display of Marine Finds." *Conservation of Marine Archaeological Objects* C. Pearson, ed., (1987): 268–270. [ECU-6]

Johnson, Jessica. "Conservation and Archaeology in Great Britain and the United States: A Comparison" *Journal of the American Institute for Conservation*. Vol. 32 No. 3, 1993. [ECU-622]

Johnson, Stuart. "Alternatives for the Continued Preservation of the U.S.S. *Cairo*." (Office of Cultural Resources Monograph, National Park Service, Southeast Region) [n.d.] [ECU-0378]

Jokilehto, Jukka. History of Architectural Conservation. London: Elsevier, 2002.

Kadry, Ahmed. "The Solar Boat of Cheops." *International Journal of Nautical Archaeology* 15 (1986): 123–131. [ECU-154]

Katzev, Michael L. "Conservation of the Kyrenia Ships, 1970–71." *National Geographic Society Research Reports* (1979): 331–340.

—. "Conservation of the Kyrenia Ships, 1971–72." *National Geographic Society Research Reports* (1980): 417–426.

Katzev, Susan W. and Michael L. Katzev. "Last Harbor for the Oldest Ship." *National Geographic* 146 (1974): 618–625. [ECU-173]

Keene, Suzanne. *Digital Collection, Museums in the Information Age*. London: Elsevier, 1998.

Kernan, Michael. "Around the Mall and Beyond." *Smithsonian Magazine* Vol. 18 No. 3, 1996. [ECU-756]

Kertish, P.J. "The Conservation of Marine Artifacts for the Professional Conservator." *Heritage Australia* 2 (1983): 62–65.

Kimble-Brown, M. "A Preservative Compound for Archaeological Material." *American Antiquity* 39 (1974): 469–473. [ECU-110]

Kvaming, L.A. "The *Wasa*: Museum and Museum Exhibit," Museum 36 (1984): 75–80.

Lafontaine, Raymond. "Recommended Environmental Monitors for Museums, Archives, and Galleries." *CCI Technical Bulletin* 3 (Canadian Conservation Institute) (1980). [ECU-453]

Lafontaine, Raymond. "Silica Gel." *CCI Technical Bulletin* 10 (Canadian Conservation Institute) (1984). [ECU-450]

Lafontaine, Raymond H. and Patricia A. Wood. "Fluorescent Lamps," *CCI Technical Bulletin* 7 (Canadian Conservation Institute) (1982). [ECU-439]

Lawson, E. "In-between: The Care of Artifacts from the Seabed in the Conservation Laboratory and Some Reasons Why it is Necessary." *Beneath the Waters of Time* (Proceedings of the Ninth Conference on Underwater Archaeology. Texas Antiquities Publication No. 6) (1978): 69–91.

Legacy. University of South Carolina. Vol. 6 No. 1, July 2001. [ECU-596]

Leigh, David. *First Aid for Finds*. University of South Hampton Press, 1972.

Logan, Judith A. and Maureen Williams. "Conservation of Inorganic Archaeological Material." *Proceedings* (Canadian Conservation Institute Seminar for Association Museums, New Brunswick Inc., June 25–26) (1991): 1–44. [ECU-373]

MacLeod, K.J. "Relative Humidity: Its Importance, Measurement and Control in Museums." *CCI Technical Bulletin* 1 (Canadian Conservation Institute) (1978). [ECU-452]

MacLeod, K.J. "Museum Lighting." *CCI Technical Bulletin* 2 (Canadian Conservation Institute) (1978). [ECU-454]

MacLeod, Ian Donald. "Environmental Effects on Shipwreck Material from Analysis of Marine Concretions." *Archaeometry: An Australian Perspective* W. Ambrose and P. Duerdin, eds., (1982): 361–367.

Maekawa, Shin and Kerstin Elert. *The Use of Oxygen-Free Environments in the Control of Museum Insect-Pests.* (Review) Getty Conservation Institute. [ECU-597]

Maming, Sam. "The *Alvin Clark* Revisited." *Wooden Boat* (1983): 66–68. [ECU-266]

Maryland Historical Trust Binder. [ECU-607]

McCarthy, Mike. "The S.S. *Xantho* Project: Management and Conservation." *Proceedings*. (ICOM, Fremantle) (1987): 9–15. [ECU-195]

—. "S.S. *Xantho*: The Pre-Disturbance, Assessment, Excavation, and Management of an Iron Steam Shipwreck off the Coast of Western Australia." *International Journal of Nautical Archaeology* (1988): 339347. [ECU-128]

McCawley, I.C. "Waterlogged Artifacts: The Challenge for Conservation." *Journal of the Canadian Conservation Institute* (1977): 17–26. [ECU-580]

McCawley, J.C., and T. G. Stone. "A Mobile Conservation Laboratory Service." *Studies in Conservation* (1982): 97–106. [ECU-245]

McCrady, Ellen. "Indoor Environment Standards: A Report on th e NYU Symposium." *Abbey Newsletter* Vol. 19 No. 7, 1995. [ECU-757]

McDonald, S. "Sharing Museum Skills at the Maritime Museum, Greenwich." *Scottish Society for Conservation and Restoration* Vol. 13 No. 2, pp. 13–14. [ECU-758]

McGrath, H.T. Jr. "The Eventual Preservation and Stabilization of the USS *Cairo*." *International Journal of Nautical Archaeology* 10 (1981): 79–94. [ECU-162]

Mecklenburg, Marion and Charles Tumosa. "Temperature and Relative Humidity Effects on the Mechanical and Chemical Stability of Collections." *ASHRAE Journal* Vol. 41 No. 4, 1999, pp. 77–82. [ECU-759]

Michalski, Stefan. "Setting Standards for Conservation : New Temperature and Relative Humidity Guidelines are Now Published." *CCI Newsletter* (Canadian Conservation Institute) No. 24, 1999. [ECU-623]

Middleton, Andrew and Janet Lang. *Radiography of Cultural Material*. London: Elsevier, 1997.

Moon, Thomas, Michael R. Schilling, and Sally Thirkettle. "A Note on the Use of False-Color Infrared Photography in Conservation." *Studies in Conservation* Vol. 37 No. 1, 1992, pp. 42–52. [ECU-653]

Muhlethaler, B., Barkman, L., and Noack. Conservation of Waterlogged Wood and Wet Leather. Paris: Edition eyrothes (1973). [ECU-0369]

Muhly, J.D., et al. "The Cape Gelidonya Shipwreck and the Bronze Age Metal Trade in the Eastern Medditerranean." *Journal of Field Archaeology* 4 (1971): 353–362.

Murdock, L.D. and T. Daley. "Polysulfide Rubber and its Application for Recording Archaeological Ship's Features in a Marine Enviroment." *International Journal of Nautical Archaeology* 10 (1981): 337–342. [ECU-163]

—. "Progress Report on the Use of FMC Polysulfied Rubber Compound for Recording Archaeological Ship's Features in a Marine Environment." *International Journal of Nautical Archaeology* 11 (1982): 349–353. [ECU-160]

Murdock, Lorne D., and John Stewart. *Monitoring Program for Shipwrecks at Fathom Five National Marine Park, Tobermoy, Ontario, Canada.* Historic Resource Conservation Branch, Parks Canada, 1800 Walkway Road, Ottawa, Ontario, Canada, KIA OM5. [ECU-523]

Murray, Howard. "The Conservation of Artifacts from the *Mary Rose*." *ICOM-WWWG Proceedings* (Ottawa) (1981): 3–18.

—. "The Conservation of Artifacts from the *Mary Rose*." *ICOM-WWWG Proceedings* (1981): 12–18. [ECU-58]

Nasti, Atilio. "Recovery and Conservation of Navigational Instruments from the Spanish Troopship *Salvador* Which Sank in 1812 in Maldonado Bay, Punta del Este, Uraguay." *International Journal of Nautical Archaeology* Vol. 32 No. 2, 2001, pp. 279–281. [ECU-624]

Nelson, Daniel A. "An Approach to the Conservation of Intact Ships Found in Deep Water." *ICOM-WWWG Proceedings* (Ottawa) (1981): 29–32. [ECU-59]

Neumann, Thomas W. and Robert M. Stanford. "Cleaning Artifacts with Calgon." *American Antiquity*. Vol. 63 No. 1, 1998. [ECU-654]

"Nitrogen" *Smithsonian* Vol. 28 No. 7, 1997, pp. 70–78. [ECU-760]

"Observations on Preservation of Archaeological Wrecks of Metal in Marine Environments." *International Journal of Nautical Archaeology* 10.1 (1981): 3–14. [ECU-418]

O'Donnell, E.B. *Oceans of Australia: Papers from the First Southern Hemisphere Conference on Maritime Archaeology* (Australia) (1978): 1–160.

—. "Conservation of Marine Artifacts from the HMS *Culloden*." *Proceedings* (13th Conference on Underwater Archaeology) (1982): 12. [ECU-65]

Oddy, W.A. and P.C. Van Geersdaele. "Lifting and Removal." *The Graveny Boat: A Tenth Century Find from Kent* (V. Fenwick, ed., British Archaeological Reports, Oxford) (1975).

Oddy, Andrew. "Restoring Faith." *Chemistry in Britain*. Vol. 31 No. 8, 1995. [ECU-655]

Olive, J. and Colin Peterson. "Conservation in Archaeology and the Applied Arts." *Proceedings of the 1975 Stockholm Conference*. IIC London, 1975. [ECU-564]

"On Site Conservation Requirements for Marine Archaeological Excavations." *International Journal of Nautical Archaeology* 6 (1977): 37–46.

"Oxygen, The Great Destroyer." *Natural History* Vol. 101 No. 8, 1992, pp. 46. [ECU-761]

Padfield, Tim, David Erhardt, and Walter Hopwood. "Trouble in Store." *IIC Washington Congress Proceedings*, 1982. [ECU-762]

Parker, Henry S. "The Chemistry and Origin of Ocean Waters." *Exploring the Oceans: An Introduction for the Traveler, Amateur, and Naturalist.* Englewood, NJ: Prentice Hall, 1985, pp. 82–106. [ECU-532]

Parker, Henry S. *Exploring the Oceans: An Introduction for the Traveler and Amateur Naturalist.* Englewood Cliffs, NJ: Prentice Hall Encyclopedia, 1985. [ECU-598]

"Parylene — A Gas Phase Polymer for Conservation of Historical Material." Nova Tron Corporation, Clear Lake, WI. [ECU-416]

"The Parylene Press—Issue 27, Winter 1997" [web site online] Specialty Coating Systems. Available at *http://www.scscookson.com/applications/winter97/*. [ECU-599]

Paterson, E.T., S.M. Mahoney, and K.E. Priydowicz. "The Conservation of the HD-4: Alexander Graham Bell's Hydrofoil." *Studies in Conservation* 24 (1979): 93–107. [ECU-180]

Pearson, Colin. "Conservation and Maritime Archaeology." *International Journal of Nautical Archaeology* 9 (1980): 147–150.

—. "The Western Australia Museum Conservation Laboratory for Marine Archaeological Material." *International Journal of Nautical Archaeology* 3 (1974): 295–305.

—. "On-Site Conservation Requirements for Marine Archaeological Excavations." *International Journal of Nautical Archaeology* 6 (1977): 37–46. [ECU-71]

—. "The State of the Art and Science of Conservation in Maritime Archaeology." *Proceedings* (First Southern Hemisphere Conference on Maritime Archaeology) (1978): 116–117.

—. "Laboratory Conservation." *Protection of the Underwater Heritage* (Technical Handbooks for Museums and Monuments, No. 4, UNESCO, Paris) (1981): 112–113.

—. "Conservation of the Underwater Heritage." *Protection of the Underwater Heritage* (Technical Handbooks for Museums and Monuments, No. 4, UNESCO, Paris) (1981). [ECU-0007]

Pearson, Colin, ed. *Conservation of Marine Archaeological Objects*, London: Butterworths Series in Conservation and Museology, 1987. [ECU-6]

Pennec, Stephane, Noel Lacoudre, and Jacques Montlucan. "The Conservation of *Titanic* Artifacts." *Bulletin* (Australian Institute for Maritime Archaeology) 13 (1989): 23–26.

Peterson, M. "The Condition of Materials Found in Salt Water." *Diving Into the Past* (Minnesota Historical Society: St. Paul) (1964).

—. "Preservation of Material Recovered from Underwater." (Office of the Chief, Naval Operations, Washington, D.C.) (1965).

Petruska-Young, Lisa. "Conservation on a Shoestring: Doing More with Less." Mid-Atlantic Conference—Ocean City, MD, 1994. [ECU-557]

Pipes, Marie-Lorraine. "Zooarchaeology: The Analysis and Interpretation of Faunal Remains from Archaeological Sites." [ECU-600]

Piercy, Robin, C.M. "Mombasa Wreck Excavation: Second Preliminary Report, 1978." *International Journal of Nautical Archaeology* 7 (1978): 301–319. [ECU-169]

Plenderleith, Harold J. "A History of Conservation." *Studies in Conservation*. Vol. 43 No. 3, 1998. [ECU-625]

Plenderleith, H.J., and A.E.A. Werner. *The Conservation of Antiques and Works of Art*. Second edition. London: Oxford University Press, 1971. [ECU-8]

Powell, Kathleen, and Patricia Wilkie. "Labeling and Tagging for Artifact Identity Survival." *Proceedings* (Pacific Northwest Conference) Vol. 1 (1976): 81–90. [ECU-233]

Pournou, A., A.M. Jones, S.T. Moss. "Biodeterioration Dynamics of Marine Wreck-Sites Determine the Need for Their In-Situ Protection." *International Journal of Nautical Archaeology* Vol. 30 No. 2, 2001, pp. 299–305. [ECU-626]

Price, Stanley, M.P., ed. *Conservation on Archaeological Excavations with Particular Reference to the Mediterranean Area*. (International Center for the Study of the Preservation and the Restoration of Cultural Deposits, Rome) (1984). [ECU-368]

Purdy, Barbara. "The Key Marco, Florida Collection: Experiment and Reflection." *American Antiquity* 39 (1974): 104–109. [ECU-108]

Purdy, Barbara A., ed. *Wet Site Archaeology*. Telford Press, West Caldwell, 1988.

"Remember the *Maine*" *Smithsonian* Vol. 28 No. 11, 1998, pp. 46–57. [ECU-763]

Robinson, Wendy S. *First Aid for Marine Finds*. (Handbook for Marine Archaeology No. 2., National Maritime Museum, Greenwich, U.K.) (1981). [ECUl-296]

—. "Observations on the Preservation of Archaeological Wrecks and Metal in Marine Environments." *International Journal of Nautical Archaeology* 10 (1981): 3–14 [ECU-78]

Rodgers, Bradley A. *The East Carolina University Conservator's Cookbook: A Methodological Approach to the Conservation of Water Soaked Artifacts*. Herbert Pascal Memorial Fund Publication, East Carolina University, Program in Maritime History and Underwater Research, 1995. [ECU-402]

—. *Conservation of Water Soaked Materials Bibliography*. 3rd ed. Herbert Pascal Memorial Fund Publication, East Carolina University, Program in maritime History and Underwater research, 1992.

Rose, Carolyn L. "Notes on Archaeological Conservation." *Bulletin of the American Institute for Conservation* 14/2 (1974): 123–130. [ECU-331]

Rule, Margaret. "Legacy From the Deep: Henry VIII's Lost Warship." *National Geographic* 163 (1983): 636–675. [ECU-0272]

Rutherford, Allison. "Infrared Photography Using an Ultravoilet Radiation Source." *Medical and Biological Illustration* 27 (1977): 35–36. [ECU-521]

Sanford, E. "Conservation of Artifacts: A Question of Survival." *Historical Archaeology* 9 (1975): 55–64. [ECU-332]

Sargent Welch VWR Scientific. Product Catalog. [ECU-602]

"Saving Grace" *Museum News* Vol. 71, 1992, pp. 42–47, 78. [ECU-764]

Schneider, Kent A. "X-Ray Radiography in Archaeology." Southeastern Archaeological Conference, Oct. 1974. [ECU-432]

Sease, Catherine. "The Case Against Using Soluable Nylon in Conservation Work." *Studies in Conservation* 26 (1981): 102–110. [ECU-221]

—. *A Conservation Manual for the Field Archaeologist*. (Archaeological Research Tools Vol. 4, Institute of Archaeology, University of California, Los Angeles) (1987). [ECU-370]

Schlichting, Carl. "Working with Polyethylene Foam and Fluted Plastic Sheet." *CCI Technical Bulletin 14* (Canadian Conservation Institute) (1994). [ECU-499]

Schneider, Kent A. "X-Ray Radiography in Archeology." (Unpublished paper presented to the Southeastern Archeological Conference, October 1974) (1974). [ECU-432]

Shapiro, L. Dennis. "Computerization of Collection Records." *Technology and Conservation*. Vol. 12, No. 3 (Fall 1995): pp. 15–19.

Shashoua, Y., S.M. Bradley, and V.D. Daniels. "Degradation of Cellulose Nitrate Adhesive." *Studies in Conservation*. Vol. 37 No. 2, 1992. [ECU-627]

Sheppard, John C. "Influence of Contaminants and Preservatives on Radiocarbon Dates." *Proceedings* (Pacific Northwest Conference) Vol. 1 (1976): 91–96. [ECU-232]

Singley, Katherine R. "Caring for Artifacts after Excavation ... Some Advice for Archaeologists." *Historical Archaeology* 15:1 (1981): 36–48. [ECU-381]

—. *The Conservation of Archaeological Artifacts from Freshwater Environments*. South Haven, Michigan: Lake Michigan Maritime Museum, 1988. [ECU-10]

—. "The Recovery and Conservation of the Brown's Ferry Vessel." *ICOM-WWWG Proceedings* (1981): 57–60. [ECU-86]

—. "Archaeological Conservation: Specialized Techniques and Research for Wet Objects." *CCI Newsletter* (Canadian Conservation Institute) No. 23, 1999. [ECU-628]

Siver, N. "Preparations for Field Conservation." *Scottish Society for Conservation and Restoration* Vol. 13 No. 2. [ECU-765]

Smith, Clifford. "Santa Lucia Conservation Complete." *Maritimes* Vol. 14 No. 1, 2001. [ECU-657]

Smith, C. Wayne. *Archaeological Conservation Using Polymers Practical Applications for Organic Artifact Stabilization*. College Station: Texas A & M University Press, 2003.

Smith, James B. and John P. Ellis. "The Preservation of Underwater Archaeological Specimens in Plastic." *Curator* 6 (1963): 32–36. [ECU-87]

Smith, R.D. "The Relative Hazards of Vapor from Solvents Used by Conservators." *Guild of Bookworkers Journal* 11 (Winter 1972–73): 3–9. [ECU-359]

"Solving Museum Insect Problems: Chemical Control." *CCI Technical Bulletin* (15, Canadian Conservation Institute) . [ECU-434]

South, Stanley A. "Notes on the Treatment Methods for the Preservation of Iron and Wooden Objects." (North Carolina Division of Archives and History) [n.d.]. [ECU-88]

Spectre, P.H. "The *Alvin Clark*: The Challenge of the Challenge." *Woodenboat* 52 (1983): 59–65. [ECU-265]

Speier, R.F.G. "Ultrasonic Cleaning of Artifacts: A Preliminary Consideration." *American* 26 (1961): 410–414.

Spriggs, J. "The Recovery and Storage of Materials from Waterlogged Deposits at York." *The Conservator* 4 (1980): 19–24.

Stavrolakes, Niki. *The Archaeology of Ships*. New York: Henry Z. Walch, 1974.

Stevens, Willis E. "The Excavation of a Mid-Sixteenth Century Basque Whaler in Red Bay, Labrador." *ICOM-WWWG Proceedings* (Ottawa) (1981): 33–37. [ECU-90]

Stewart, John, Lorne Murdock, and Peter Waddell. *Reburial of the Red Bay Wreck as a Form of Preservation and Protection of the Historic Resource*. Historic Resource Conservation Branch, Parks Canada, Ottawa, Ontario, Canada, KIA OM5. [ECU-524]

Strang, Thomas J.K., and John E. Dawson. "Controlling Museum Fungal Problems." *CCI Technical Bulletin* (No. 12, Canadian Conservation Institute, Ottawa) (1991). [ECU-371]

Stoffyn-Eglit, Patricia and Dale E. Buckley. "The Micro-World of the *Titanic*." *Chemistry in Britain* Vol. 31 No. 7, 1995. [ECU-658]

Suenson-Taylor, Kirsten, Dean Sully, and Clive Orton. "Data in Conservation: the Missing Link in the Process." *Studies in Conservation* Vol. 44 No. 3, 1999, pp. 184–194. [ECU-629]

Switzer, R.R. "Some Technical Observations Concerning the Preservation of the *Bertrand*." (National Park Service) (1969).

Technology and Conservation Magazine. *Technology and Conservation Magazine* 4/93 and 2/94. Winter 1993 and Summer 1994. [ECU-520]

Thomson, Gary. *Museum Environment*. London: Butterworths, 1986.

Thorne, Robert M. "Site Stabilization Information Sources." *Technical Brief* (National Park Service Cultural Resources, No. 12) (1991): 1–8. [ECU-374]

Titford, Emma. "The Importance of Packing Objects." *British Maritime Museum Quarterly* 2 (1989): 28–29. [ECU-92]

—. "The Conservation of Artifacts from the New Old Spaniard." *British Maritime Museum Quarterly* 2 (1989): 25–27. [ECU-91]

Townsend, S. "The Conservation of Artifacts from Salt Water." *Diving Into the Past* (Minnesota Historical Society: St. Paul) (1964).

—. "Standard Conservation Procedures." *Underwater Archaeology, A Nascent Discipline* (UNESCO, Paris) (1972): 251–256.

Tuck, James A. "Wet Sites Archaeology at Red Bay, Labrador." *Wet Site Archaeology* B.A. Purdy, ed., (1988): 103–111.

Turner, W. "Appendix 3: Conservation of the Finds in Piercy, R., Mombasa Wreck Excavation: Second Preliminary Report, 1978." *International Journal of Nautical Archaeology* 7 (1978): 317–319.

Tylecote, R.F. "Durable Materials for Sea Water: The Archaeological Evidence." *International Journal of Nautical Archaeology* 6 (1977): 269–283. [ECU-161]

"Underwater Treasures Under the Microscope: Objects from the *DeBrack* Analyzed at Winterthur." *Winterthur Newsletter* 34 [n.d.]: 14–15.

Wachsman, Shelley, K. Raveh, and O. Cohen. "The Kenneret Boat Project, Part I. The Excavation and Conservation of the Kenneret Boat." *International Journal of Nautical Archaeology* 16 (1987): 233–245. [ECU-129]

Wainwright, Ian N.M. "The Science of Conservation : CCI's New Variable—Pressure Scanning Electron Microscope with X-Ray." *CCI Newsletter* (Canadian Conservation Institute) No. 29, 2002. [ECU-630]

Walston, S. "Conservation in Australia." *Proceedings* (ICCM National Conference, Canberra, Institute for the Conservation of Cultural Material, The Australian Museum, Sydney) (1976).

Watkinson, David, ed. *First Aid for Finds* (United Kingdom Institute for Conservation, Rescue-The British Archaeological Trust, London) (1987). [ECU-429]

Wedinger, Robert. "Preserving our Written Heritage." *Chemistry in Britain* Vol. 28 No. 10, 1992. [ECU-659]

Weier, L.E. "The Deterioration of Inorganic Material Under the Sea." *Bulletin* (Institute of Archaeology, University of London II) (1973): 131–136.

Werner, A.E.A. "Preservation Techniques." *Education in Chemistry* 9 (1972): 131–133.

—. "New Materials in the Conservation of Antiquities." *Museums Journal* 64 (1972): 5–15. [ECU-404]

"What is 'Museum Quality'? A Curator Looks at Boat Restoration." *Woodenboat* Vol. 144, 1998, pp. 84–90. [ECU-766]

Winter, J. ed. "ICOM Reports on Technical Studies and Conservation." *Art and Archaeology Technical Abstracts* 14, No. 2, [n.d.] [ECU-328]

Wright, M.M. "The Conservation Laboratories at Marischal Museum, Aberdeen." *Scottish Society for Conservation and Restoration* Vol. 7 No. 2, 1996. [ECU-798]

Young, Gregory. "Impact of Climatic Change on Conservation." *CCI Newsletter* (Canadian Conservation Institute) No. 26, 2000. [ECU-631]

Young, Janie Chester, "The State of the Art and the Artifact: A Regional Survey of Museum Collections," *Technology and Conservation Magazine*. Vol. 11 Nos. 2–3 (Summer–Fall 1992): 10–16.

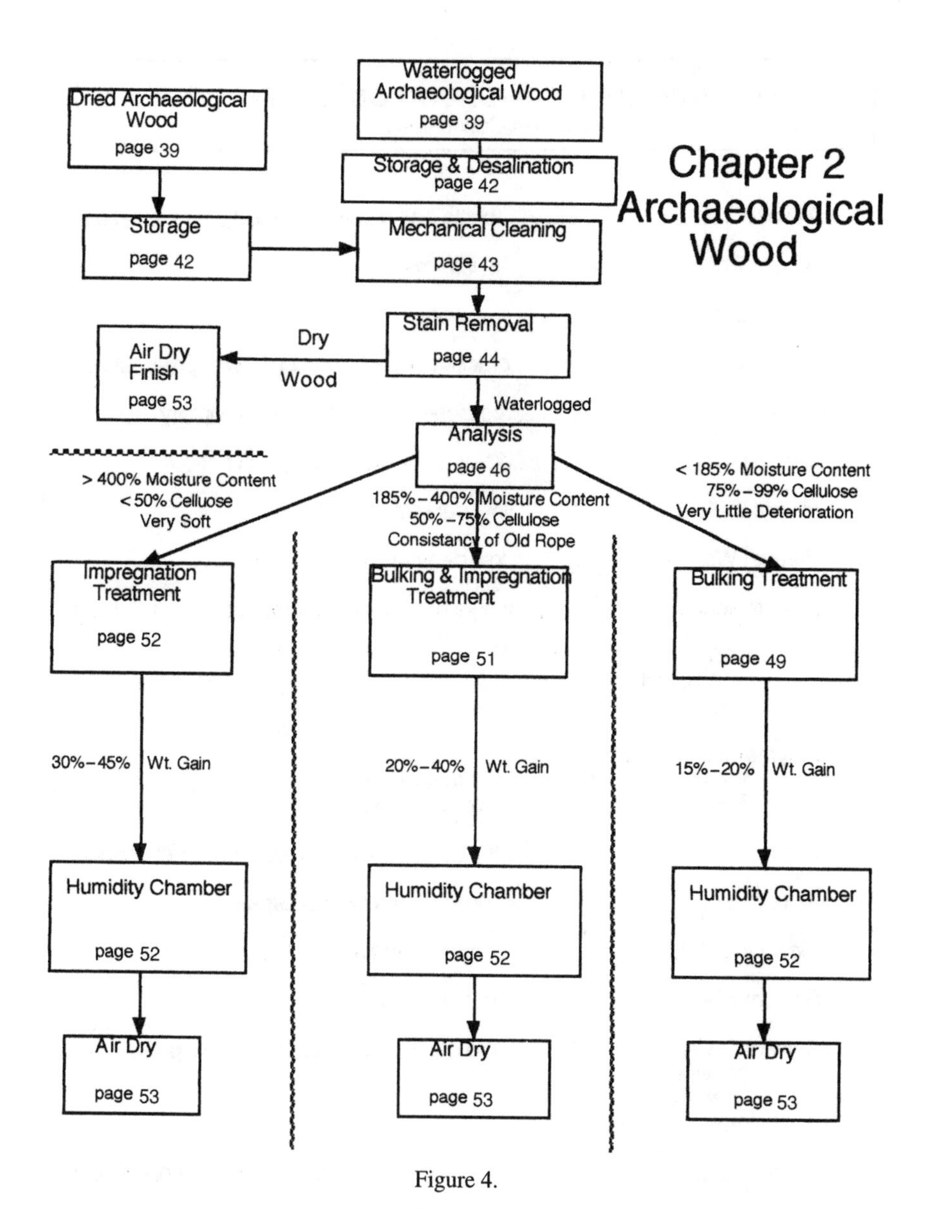

Dried Archaeological Wood
page 39
Waterlogged Archaeological Wood
page 39
Chapter 2
Archaeological Wood
Storage & Desalination
page 42
Storage
page 42
Mechanical Cleaning
page 43
Stain Removal
page 44
Dry
Wood
Air Dry Finish
page 53
Waterlogged
Analysis
page 46
> 400% Moisture Content
< 50% Celluose
Very Soft
185% – 400% Moisture Content
50% – 75% Cellulose
Consistancy of Old Rope
< 185% Moisture Content
75% – 99% Cellulose
Very Little Deterioration
Impregnation Treatment
page 52
Bulking & Impregnation Treatment
page 51
Bulking Treatment
page 49
30% – 45%
Wt. Gain
20% – 40%
Wt. Gain
15% – 20%
Wt. Gain
Humidity Chamber
page 52
Humidity Chamber
page 52
Humidity Chamber
page 52
Air Dry
page 53
Air Dry
page 53
Air Dry
page 53

Figure 4.

Treatments for Archaeological Wood
Represented in Conservation Liturature

Alum..brittle and heavy, not good dimentional control

Barium Borate...........................no information

Creosote..................................carcinogen

Freeze Dry................................cracks cell walls, must use with pre-treatment

Glycerol.....................................collapse, shrinkage, hygroscopicity

Lindseed Oil.............................elasticity, allows moisture in not out

Methyl Cellulose.......................low penetration

Paraffin Wax.............................low penetration, dark and heavy

Polyethylene Glycol................works well in both Impregnation and Bulking
 400
 600
 540 Blend
 4000

Polymerization of a Monomer
 Vynyl Monomer.............not reversible
 Silicon Oil.........................not reversible

Rosin..fire hazard, changes appearance of artifact

Salt..does not prevent collapse

Slow Dry.....................................does not prevent collapse

Sodium Silicate...........................not reversible

Solvent Drying............................fire hazard, do not prevent collapse
 ethanol
 acetone
 methanol

Sugars..all work as bulking agents with added heat
 fructose
 manitol
 sorbitol
 sucrose
 unrefined

Tetraethoxy Silane (TEOS)........not reversible

Figure 5.

Chapter 2

Archaeological Wood

WOOD FORMATION AND DEGRADATION—THEORY

Wood technology and tool making has surpassed virtually all other human worked materials in variety of use and multiplicity of properties. On many archaeological sites, wood, or wooden remains, make up most of the primary artifacts, hence its inclusion in this manual as the first material type. Wood is a durable easily worked organic material produced by trees and shrubs. Human tool manufacturing in wood likely pre-dates culture and language. It was used to make a multitude of diverse artifacts from weapons, homes, and shelters, to musical instruments, shoes, and ships. But in order to understand wood use in tools and artifacts it is necessary to understand some of wood's characteristics as well as its make-up.

Wood, like some modern materials such as plastic, has many different traits and each wood type had its historical use. For example, white oak is strong, durable, and reasonably waterproof but its too brittle make a good bow or handle for a digging tool and lacks the flexibility of yew or ash. Elm is renown for its mallet head endurance and yellow and white pines were easily worked and flexible but not exceedingly durable or strong. Other woods such as cedar are both strong and long lasting in a variety of uses. Pre-modern people were taught the value of each type of wood according to cultural need. Therefore, in all its variety of species, wood cannot truly be looked upon as a homogenous substance, but rather as a group of materials with varying properties used to produce a myriad of artifacts.

A good deal of research has been conducted on archaeological wood. Some of the classic inquiries include H. J. Plenderleith and A.E.A. Werner's, *The Conservation of Antiquities and Works of Art*, 1956, or A. J. Stamm's earlier work on the search for a water soluble wax. Other articles include Brorson Christensen and J. de Jong's prolonged search for methods to classify the degradation of archaeological wood. Another classic reference found in most museums is, B. Muhlenthaaler, Lars Barkman, and Noack's *Conservation of Waterlogged Wood and Wet Leather*, 1973. Yet archaeologists and conservators are not the only people interested in archaeological wood. Until fairly recently wood anatomy and taxonomy have been the exclusive purview of biologists and wood scientists who have also produced good reference works such as A.J. Panshin and C. de Zeeuw's, *Textbook of Wood Technology*, 1970, which is a very useful aid in the microscopic identification of wood species.

Wood taxonomy breaks down all species into three categories, monocotyledons, hardwoods (dicotyledons and angiosperms), and softwoods (gymnosperms and coniferals). Most archaeological artifacts are made of hardwoods or softwoods, though monocotyledons such as palm and bamboo were used in the tropics. Hardwoods include birch, oak, elm, beech, cherry, hickory, maple, willow, walnut, and sweet gum to name a few, and all of these tree types come in more than one variety. Softwoods include the many varieties of pines, fir, spruce, hemlock, redwood, cedar, and larch. Generally speaking hardwoods are more dense and complex than softwoods.

Looking at a log that has been crosscut or sectioned from the outside inward you first notice the protective bark on the outside. Just inside the bark is a thin layer, usually less than an inch in thickness called the cambrium or sapwood. This layer is the only living part of wood and is made up of phloem cells that conduct the food manufactured in the leaves to the remainder of the tree and the roots. Inside the cambrium is the heartwood made up of xylem cells that conduct water from the root to the upper extremities of the tree. The heartwood is what we are most concerned with as conservators as this is the structure of the tree that is most useful for manufacturing tools and artifacts. Radiating out from the heartwood are rays consisting of ray cells that help distribute water from the roots throughout the tree.

Inside the wood tree cells are arranged longitudinally except for the ray cells. Since the phloem is the only living part of a tree trunk the wood is successively formed of phloem cells. Each year as these cells die they are surrounded by another year's growth and move to the interior of the tree forming the heartwood. These wood cells, called tracheids, are bean pod shaped, closely bundled together, and gathered around vessels that permeate the wood. Wood cells are largely hollow. The cavity in the wood cell is known as a lumina. Early wood (spring growth) cells are larger and longer than late wood (late summer and fall) cells that tend to be smaller with thicker cell walls. Early wood appears light colored while late wood looks dark giving wood its distinctive annular rings. Softwood tracheids are up to 7mm long in early wood and 25–80 microns (1 millionth of a meter) in diameter shrinking to 5 microns by late season. Hardwood tracheids are smaller and denser at up to 1.5 mm in length and varying 1 to 10 microns in diameter depending on the season (Pearson, 1987; 57).

Wood cells are made of complex carbohydrate molecules of cellulose, hemicellulose, lignin, and holocellulose. Cellulose is by far the most important of these molecules and makes up the greatest percentage of a wood cell. Cellulose molecules join together to form micro-fibrils that in turn, create macro-fibrils that give wood cell walls their strength and durability.

Tracheid cell walls are permeable to an extent depending on the bordered pits. Pits are holes in the cell walls that allow water to pass through from one cell to the next, but only if the valve (margo and torus) in the pit is open. On dying early wood cell pits close off (aspirate) while late wood pits remain open (non-aspirated), leading to the contradiction that although late wood is denser than early wood, it is also more permeable.

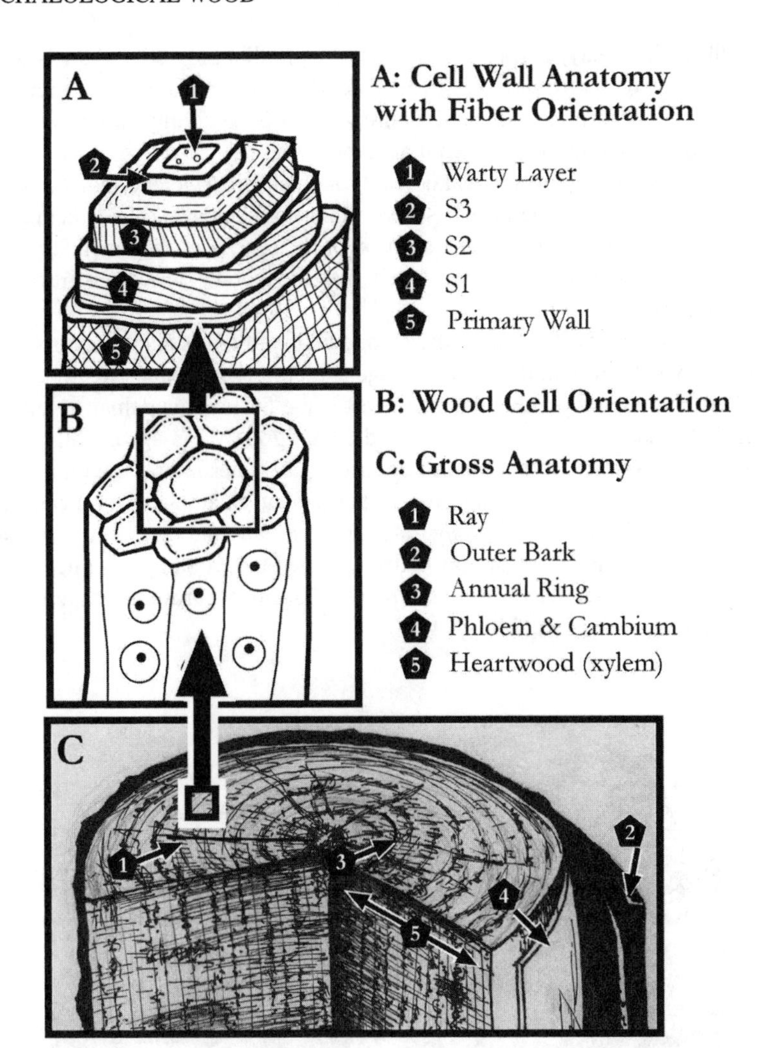

Figure 6. Wood is mainly comprised of the complex carbohydrate cellulose. Micro-fiber directions in the secondary cell wall (S1, S2, S3) are responsible for the overall drying behavior. Illustration by Nathan Richards.

Wood cells contain a primary cell wall and the secondary cell wall. Conservators are most concerned with the secondary tracheid wall as it is accountable for most of wood's strength. However, the secondary cell wall is also subject to the greatest damage.

The secondary cell wall is arranged in three distinct layers S1, S2, and S3. As can be seen in the illustration the fibril orientation of each of these layers is different as is the thickness of each layer. S2 dominates the secondary cell wall and the fibril direction spirals longitudinally around the core of the wood cell. The S1 and S3 fibrils lay nearly perpendicular to the longitudinal direction the wood cell. The orientation of the fibrils is key to how wood expands and contracts with water absorption and desorption. Since S2 dominates, wood generally expands more in cross section than lengthwise when it absorbs water.

Cell walls are subject to breakdown through physical weathering, the actions of micro-organisms, or chemical dissolution through acids, alkaloids, and pollutants. In fact, the structure and preservation of the largest of the secondary wood cell layers, S2, determines to a great extent how the wood in any given artifact has held up over time.

The orientation of the secondary wood cell wall fibers also determines that damage to wood will manifest in shrinkage and collapse in three directions. Tangential and radial shrinkage will be greatest if the S2 layer is damaged and longitudinal shrinkage will manifest if S1 and S3 are greatly damaged. Collapse and shrinkage are technical terms used to describe how water interacts with archaeological wood. The water logging process permanently damages wood causing it to collapse and shrink when it dehydrates. The amount and direction of collapse and shrinkage are indicators of how badly damaged the wood is on a microscopic level and which layers have sustained the greatest impact from the waterlogging process (see section on waterlogging).

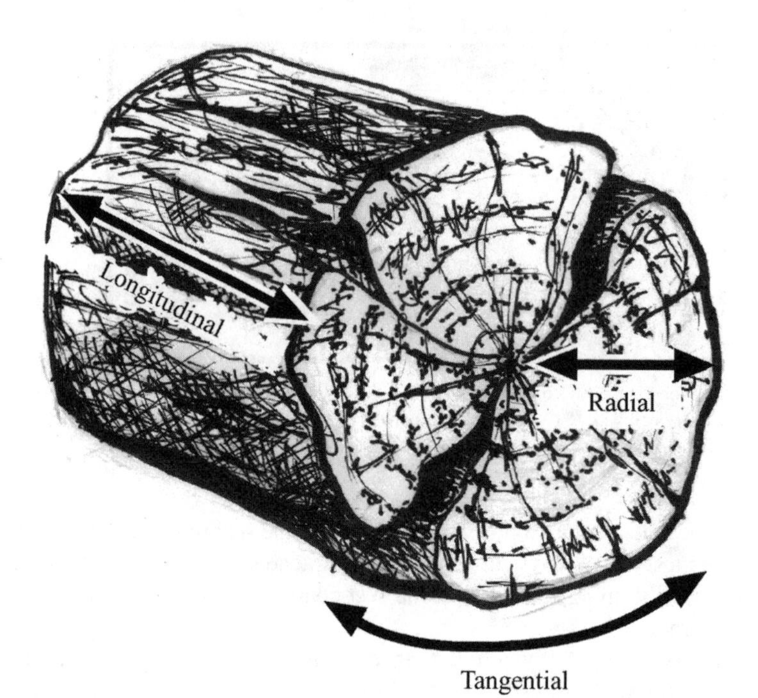

Figure 7. The three directions of collapse and shrinkage in Archaeological Wood. Illustration by Nathan Richards.

ARCHAEOLOGICAL WOOD—THEORY

Archaeological wood can be highly degraded or may be no different than the stout objects we use or come into contact with everyday. Wooden artifacts are obviously much older than the sort we deal with daily and may show their age in the form of structural weakness, cracking, discoloration, face checking, or a weathered appearance. In most instances archaeological wood has been attacked by living organisms on both a microscopic and macroscopic level, causing a weakening in its structure and possibly the destruction of its outer surface. Bacteria and fungi living in the ground will digest the complex carbohydrates that make up the wood and it will decay depending largely on the moisture content and drainage of the soil. Other creatures such as termites can quickly destroy and consume artifacts made of wood.

Various chemical pollutants and acids can also break down wood lying in the soil while acidic metallic salts from corroding metal may break down wood at a cellular level and even mimic the shape of wood cells that it has replaced. Still other wooden artifacts interred below the water table, in lakes, rivers, or oceans, are likely to be waterlogged. Dry, waterlogged, wet, or even damp wood recovered from an archaeological site will need the attention of the Minimal Intervention Laboratory staff. It cannot be assumed by appearance alone that wood is stable and needs no further conservation treatment. The inclusion of metal fasteners, stains, and dirt likely mean that some form of conservation intervention must be considered.

WATERLOGGED WOOD THEORY

Waterlogged wood can be defined as wood that contains little or no interstitial air in its cellular spaces, capillaries and micro-capillaries. It has also been altered (weakened) by the micro-degradation of the cell walls. Waterlogged wood demonstrates a moisture content (MC) at or above the fiber saturation point (FSP) or that point at which all wood fibers and micro-fibrils in the cell walls are saturated with water. On dehydration waterlogged wood can exhibit cellular collapse at or above the FSP as water rushes out of the wood cells one at a time. The sudden rush of water from the interior of the cell is not easily described and best seen through a microscope. Since water exhibits a high cohesiveness (capillary tension) it pulls the wood cell walls inward as it rushes out of a cell collapsing the wood cell in a microscopic, yet violent act. Collapse is easily observable with the naked eye in waterlogged archaeological wood that is allowed to dry, as fissures and cracks appear suddenly in the structure of the wood. Even wet or damp waterlogged wood will collapse if it is allowed to rest in the atmosphere without constant dousing with water. Capillary tension collapse is a permanent condition in dehydrated waterlogged wood. Wood cells, even on rehydration will not expand to their previous size.

Further dehydration below the fibre saturation point will lead to Shrinkage of the cell walls and further distortion of dried waterlogged wood. Shrinkage, unlike collapse, is to some extent reversible on re-hydration. It can be defined as the loss of water in the cell wall through evaporation. This water is normally locked into the cellulosic fibrils but can evaporate in dry conditions.

Green wood or unseasoned wood, even with a considerable water content, contain air pockets within the tracheids. These air pockets expand on dehydration allowing water to leave green wood cells without causing the tremendous capillary tension pressure responsible for the collapse of waterlogged wood. Green wood does exhibit some reversible shrinkage of the cell walls on complete or near dehydration and both green and waterlogged wood swell 8% to 12% when fully saturated depending on specie.

IMPREGNATION AND BULKING—THEORY

Capillary tension collapse and shrinkage complicate the process of treating waterlogged wooden artifacts. It is easy to see in the flow chart on page 60 that the procedure for treating waterlogged wood is far more complex than that used to treat dry recovered archaeological wood. Conservators treating waterlogged wood attempt to arrest or mitigate collapse and shrinkage and add strength to the artifact, to maintain its dimensions and appearance.

There are essentially two methods for conserving waterlogged wood, impregnation and bulking. Impregnation is defined as the filling of all wood spaces and wood cells with an inert material that hardens upon drying in order to add strength to the wood and at the same time reduce capillary tension collapse. Collapse is reduced when the impregnating agent demonstrates less surface cohesion than water. Impregnation is reserved for waterlogged wood that is in very bad condition, where much of the cellulose in the cell walls has been destroyed and removed. Impregnating agents include polyethylene glycol (PEG) of molecular weight greater than 1000, alum, rosin, barium borate, and monomers that can be polymerized within the wood.

Bulking is a less drastic procedure than impregnation. It can be defined as the insertion of an inert material into the cell walls to replace missing cellulose in order to strengthen the wood cell walls and make them more resistant to collapse and shrinkage. As with impregnating agents, bulking agents reduce collapse by lessening water's surface cohesion. Bulking is used on waterlogged wood that has little to moderate microscopic damage and still retains much of its cellulose. Bulking agents include PEG with molecular weight less than 1000, various sugars, salts, or vinyl monomers.

Bulking and impregnation are proven and highly successful treatments in the conservation of waterlogged wood, the difficulty for the conservator lies in choosing which treatment to perform. In some instances the choice is easy, highly degraded wood is impregnated and very lightly degraded wood is bulked. Fortunately there are objective tests and analysis available to help make this decision.

Wood analysis is a continuing theme is waterlogged wood conservation research. Pioneers such as Lars Barkman, Moren and Centerwall, R. M. Organ and A. J. Stamm all agreed that the problem with waterlogged wood is that it is not uniform, but rather a combination of sound and degraded wood. Although these researchers agreed that analysis should determine the type of treatment, laboratory tests were not standardized. Experimental tests followed objective or subjective paradigms but for many years did nothing to solve the problem presented by waterlogged wood. Finally however, in the 1970s, J. de Jong, assigned moisture content (MC) to B. Christensen's three subjective waterlogged wood classifications, and for the first time produced a simple and useful way to analyze waterlogged wood.

Christensen Classification	deJong Classification
A. Very Soft—Little Cellulose	I. >400% MC
B. Thin Zone—Little Cellulose, Consistency of Old Rope	II. 185%–400% MC
C. Little Deterioration—Non-permeable	III. <185% MC

Within a decade, David Grattan of the Canadian Conservation Institute upgraded wood analysis by suggesting that specific gravity (SG - density measured in grams per cubic centimeter) comparisons to known speciated wood values actually provide the best indicator of the deterioration of waterlogged wood. In other words Grattan compared the weight of dried waterlogged wood samples to the known values for the same sized sample of a particular species and deduced theoretically how much wood (cellulose) was missing.

Archaeologist/conservators are the beneficiary of this important research conducted on waterlogged wood analysis. The methodological analysis section of this chapter divides waterlogged wood into Christensen and deJong's three categories using both moisture content and wood description. The manual goes on, however, to use cellulose content (a product of Grattan's specific gravity) as a third determiner. These descriptors are easily determined and should enable conservator's and technicians to divide artifacts into those to be bulked, those to be impregnated, and those that will receive both treatments.

METHODOLOGY

As this chapter moves from a theoretical look at wood taxonomy and structure to the actual methodology of conservation, some major concepts should be kept in mind. First, all artifacts are different, so treatment procedures are to some degree subjective and must be modified for each object, this is the art of conservation. Many treatments are intrinsically harmful to an artifact for a short period so treatment effects must be limited to a specific area on an object or the treatment must be discontinued when it is observed that the object is being harmed.

Simply handling an artifact can cause damage and unfortunately all artifacts must be handled for successful treatment (clean cotton gloves are always a good idea in handling artifacts undergoing treatment). Therefore, a conservator must always carefully observe an object before and during treatment and compare any possible harm to probable good. This is particularly true during the cleaning (mechanical and stain removal) and the treatment phase. There is always that point where the good done by removing the last grain of dirt, or the last indication of stain; is undone by the damage that may be brought about through handling and over aggressively treating an artifact. Acumen based on experience and common sense is a good trait in a conservator or conservation technician. At times a judgement call is necessary to dictate when a treatment is finished.

STORAGE OF DRY ARCHAEOLOGICAL WOOD—METHODOLOGY

Archaeological wood recovered from a land site can be in virtually any condition imaginable. If it appears to be in good condition and has not been recovered from below the water table it should be gently cleaned with a paint brush and wrapped in plastic and carefully boxed for its return to the archaeological laboratory. At the lab the plastic should be removed or opened to avoid water condensation build up and resulting soaked areas inside the plastic. Plastic is a durable inert non-acidic material and should always remain between the artifact and the self or box it rests in. At this point the artifact can be stored on a shelf for a fairly long period of time provided the relative humidity of the room does not exceed about 60%, Extremely low and high humidity levels must be avoided as the artifact will swell or shrink accordingly.

Obviously all artifacts recovered from an out door environment has been exposed to the elements. Periodic rainfall or even sporadic flooding of an area don't necessarily mean that the wooden artifact is waterlogged. If an artifact was waterlogged at some time in its internment, and it has been recovered dry, then it will be in a collapsed state. There is yet no process to save it from the effects of capillary tension collapse and reverse the process. In other words, the damage has been done and it must now be treated as a dry recovered artifact.

Should the artifact be wet or damp on recovery and the excavated area lays below the water table the artifact should be treated as a waterlogged object. Above all it should not be allowed to dry. Complete immersion (as opposed to a damp wrap) is recommended for its transportation back to the conservation lab.

WATERLOGGED ARCHAEOLOGICAL WOOD STORAGE—METHODOLOGY

Wood recovered from salt water environments should not be immediately immersed in fresh water. The cell walls act as semi-permeable membranes prohibiting the free mixing of the two liquids. Salinity differences between the water inside and outside the wood cells will trigger an osmotic pressure differential.

Typically the salt-water solution within the wood cells becomes hypertonic, moving outward to merge with the fresh water. Fresh water is prohibited by the cell wall from moving into the cell as fast as the salt water vacates creating capillary tension collapse. This may be dubbed the bathtub effect, as anyone who has enjoyed a long bath experiences the phenomena in the form of shriveled fingers and toes.

A gradual introduction to fresh water all but eliminates this problem. Recovered wood should be placed initially in a solution 50% fresh water and 50% salt water. The rate of subsequent fresh water introduction is dependent on the size of the artifact. The bulkier the artifact the slower the process. Generally, one week between fresh water additions is sufficient. Each addition is 50% of the artifact solution into 50% fresh water. After two subsequent fresh water additions the water salinity should be no more than four parts per thousand. At this point the artifact can be safely immerse in fresh water.

The final outcome of the fresh water immersion is the total desalination of the artifact, a very necessary step in the conservation of waterlogged wood. Yet the wood is now resting in a completely different environment than that which it had come to equilibrium with. It will begin to deteriorate at an accelerated pace. Only a few months in storage will add the equivalent of decades to the appearance of the artifact. The storage solution will also begin to be colonized with fresh water bacteria. To guard against this, regularly add 50 ml of Lysol per 10 liters of storage water and cover the storage container.

MECHANICAL CLEANING—METHODOLOGY

Handle with care is a good slogan for mechanical cleaning. The procedure used to mechanically clean dry archaeological wood is the same for cleaning waterlogged archaeological wood. Both should be treated as though they are extremely delicate, particularly the outer surface layer that may contain the archaeological, or worked surface.

Mud, and or silt removal are easily accomplished under gently running water with soft nylon bristle brushes (either paint brushes or soft tooth brushes). A dental pick can be used for small intrusions. Sometimes, depending on how much mud is attached to the artifact, this cleaning should take place outside to prevent clogging sink traps. Squirt bottles also make good cleaners and may supply all of the water necessary to accomplish the task. It should be kept in mind that the external surface of the artifact will be in the worst condition, sliding an object in the dirt and grit removed from it is tantamount to buffing it with sandpaper. At times the artifact may need to be wrapped in cloth to protect it from abrasion.

Waterlogged wooden artifacts are constantly wetted during cleaning, this is unnecessary for dry finds. Capillary tension collapse can occur above the fiber saturation point, so even if the waterlogged artifact feels and looks damp it may be undergoing distortion on a cellular level. Small irreversible cracks and fissures will be the first indication that this may be happening.

Concretions can be a problem during mechanical cleaning. Dry finds may have nail or fastener concretions consisting of iron corrosion product and sand. These can be extremely hard to deal with as the concretion itself is much harder than the wood. These concretions will require use of dental picks, scalpels and perhaps an electric or pneumatic air scribe. Scribes are small hand held devices resembling a pencil-sized jack-hammer, usually used to permanently etch materials or equipment. The vibrating action of the scribe bit can be controlled to lesson or increase the effects of the instrument.

Barnacle and oyster shells are a major impediment to mechanically cleaning waterlogged wood. Often these shells have to be removed with a forceps and scalpel. However, the shells may contour the archaeological surface making it difficult to remove them without damage to the artifact. The shell contour may be a good way to observe the pattern originally present on the wood. This is particularly true for fine cabinetry and molding, the shells should be examined before being discarded.

Mechanical cleaning is a painstaking process that requires a great deal of patience to prevent damage to an object. It will likely be the most important step in uncovering the details of an artifact's wear, usage, and fabrication, and is the first step in the micro-excavation of an artifact. Unlike other steps in the treatment and conservation of archaeological wood, mechanical cleaning can be more or less left up to the imagination. Work stations featuring lighted magnifying lenses, foam padding, running water, and an assortment of dental picks, tools, knives, and scribes can be arranged to the conservator's whim and sense of efficiency.

After mechanical cleaning the artifact is returned to storage. For dry finds this will mean air drying and a return to the storage shelf. Waterlogged artifacts are returned to fresh water storage. While the artifact is undergoing mechanical cleaning its storage container should be checked for biological growth, dirt and other detritus, scrubbed clean and its storage solution replaced. Biotic growth may be an indicator that the wood is continuing to deteriorate in storage and may indicate that it is time for another dose of biocide (see waterlogged wood storage).

STAIN REMOVAL—METHODOLOGY

Stains are not just unsightly reminders that a wooden artifact has had a long and varied history. It may also indicate that the wood is under chemical assault. Anyone who has lost cotton clothing through the effects of iron stains can attest to the fact that stains are harmful to cellulosic materials. Stains also represent substances, along with sodium carbonate concretions, that can block bordered pits in the cell wall hindering the introduction of bulking and impregnating material into wood. Removing stains and concretions may be vital to saving a wooden artifact.

Stains come in several varieties but their makeup is usually quite easily seen and categorized. Quite often stains represent metallic salts from leaching corrosion products, orange from iron, or blue-green from copper. Anaerobic (airless) breakdown of these metals will produce black and brown stains. But black stains may also represent sulphide staining from organic substances. It is up to the project archaeologist to decide if a stain represents a hint at artifact usage, or is the result of something that happened to the artifact during the site formation process. Stains that indicate use and wear, or that perhaps attest to what was contained in the artifact during its useful life should be preserved. Those accidental stains resulting from long sea bottom (benthic) or underground internment should be noted, recorded (for they indicate site formation process), processed, and removed. The difficulty here is that virtually every stain represents a foreign substance that will likely start to break down the artifact on a cellular level. For now there is no good answer concerning how to neutralize stains that complete an interpretation of an artifact, because buffering agents that would normally neutralize the affects of a stain are also harmful to wood.

If a stain is not part of the interpretation of an object and needs to be removed, there are simple procedures to do this. Iron and copper stains can be removed from the artifact with poultices of dilute acid. Acids such as citric, oxalic, or hydrofluoric can be used in concentrations of 3% to 10%. Citric is the best all purpose acid as its harmful effects are negligible and it is useful for many other procedures making it a multi-purpose chemical (both oxalic and hydrofluoric carry health concerns—see MSDS). As acids will break down the wood on a cellular level, the stain removal should be concentrated only on the stain, a general soak is not recommended. Mix a 3% solution by weighing out 3 grams of citric crystals and adding it to 97 ml of water. A poultice to concentrate the acid on the stain can be manufactured of acid soaked cotton-balls, or by mixing the acid with talcum powder and placing the paste on the stain. Waterlogged wood will of course need to remain soaked with water during this entire treatment.

The poultice will need to be changed several times to remove an average stain and treatments can last several days. A buffering solution of dilute sodium carbonate administered with a squirt bottle can be used to neutralize any acid present in the treated area. After the stain has been removed the treated area will need to be given a thorough rinse with distilled, deionized, or rain water.

Organic stains are neutralized with hydrogen peroxide. The conservator can employ a 3% to 10% hydrogen peroxide solution using the poultice method. Hydrogen peroxide is available in 3% concentration from any drug store. Again, the effects of the stain removal need to be carefully monitored and treatment discontinued if damaging effects are observed. At that point a more dilute solution can be applied or the treatment can be discontinued for the sake of the artifact.

All stain removing soaks and poultice treatments should be followed by meticulous rinses. It does no good to remove a stain (generally representing an acid intrusion), if the acid used is not fully neutralized. This rinse should last for several hours and the water changed several times.

WATERLOGGED WOOD ANALYSIS—METHOD

The following easily preformed tests are derived from the research literature (see Theoretical Section) and have demonstrated their functionality in matching wood deterioration to treatment type.

1. Moisture Content (MC).
2. Specific Gravity Determination (SG).
3. Drying behavior of a sample (DB).

It should be noted that all three tests are not preformed on every wooden artifact that comes to the laboratory, only representative samples. Testing is simple but does require the sacrifice of roughly one cubic centimeter of artifact per test. This may be no problem if the conservator is dealing with an entire wooden shipwreck, but can be problematic if the artifact under analysis is small and delicate. Analysis is yet another judgement call by the conservator who must ask, do the probable benefits outweigh the loss of a small part of an artifact? If the answer is no, a subjective description from Christensen's original research is included with the moisture content and cellulose percent to help guide the artifact into the correct process, impregnation or bulking. These subjective descriptors include very little deterioration, consistency of old rope, and very soft little cellulose. In the end, any conservator that has repeatedly conducted these tests will gain an eye for judging the moisture content if not the specific gravity of an artifact. Experience will enable the conservator or technician to guess with a high degree of certainty which of the three directions the artifact needs to travel for treatment.

1. Moisture Content

A small piece of wood must be removed from a sample artifact representing at least one cubic centimeter. A scalpel, knife or sometime a saw must be used for this purpose. If there is a difference in the wood conditions between the inner and outer layers, both should be sampled. If there is enough wood, an increment borer will sample both the outer and inner wood enabling the conservator to observe any differences. If the outer wood is degraded significantly it is possible that the inner core is in better condition.

This sample is measured to determine its volume in cubic centimeters. If it is a cube the volume is a simple calculation of length x height x width in millimeters divided by 1000 to equal cubic centimeters. If it is an irregular piece it can be placed into a small graduated cylinder filled to an observed level with water. On immersing it into the water in the cylinder, its volume can be seen in how much water it displaces. At room temperature one milliliter (ml) is equal to one cubic centimeter (cm^3). After measurement for volume the sample is weighed on a balance beam both before and after oven dehydration. The weights can be plugged into the following formula to calculate the moisture content of the sample.

Moisture Content

$$\frac{\text{Weight Waterlogged} - \text{Weight Oven Dry}}{\text{Weight Oven Dry}} \times 100 = \text{MC}$$

This formula can give a fast appraisal of the deterioration present in a wood sample. Green wood can normally contain anywhere from 20% to 60% MC. Waterlogged archaeological wood seldom contains less than 90% MC and this figure can increase to as high as 800% or 900% MC in badly degraded waterlogged wood.

A simple example of MC calculations may start with a sample that weighs 2.7 grams. After the sample was dried in an oven it weighed .6 grams. When these figure are plugged into the formula for moisture content they reveal that the sample has a 350% MC. Therefore, the artifact from which this sample was taken is eligible for both a bulking and impregnation treatment as seen on the flow chart indicator on page 33.

Though moisture content is usually a good indication of the deterioration present in waterlogged wood it fails if the wood has already been dried and collapsed, or if the wood is not fully waterlogged. Both of these conditions will give artificially low moisture contents.

2. Specific Gravity (SG)

Specific gravity can be a good indicator of the degradation in waterlogged wood. It is a density determination always calculated in grams per cubic centimeter and varies with wood species. The following table lists some examples of specific gravity in gr/cm^3. It is easy from this table to see the density differences between hardwood and softwoods at the dotted line.

1. Water	1.00
2. Balsa	.11
3. Eastern Red Cedar	.47
4. Western White Pine	.38
5. Red Pine	.44
6. Eastern White Pine	.35
7. Southern Yellow Pine	.51–.61
....	
8. White oak	.68
9. Red oak	.63
10. American Elm	.50
11. American Chestnut	.43
12. Live Oak	.88
13. Lignum Vitae	1.14

But the specific gravity measurement is far more important archaeologically than just comparing the density of different wood species. It can be used to determine how much cellulose remains in a wood sample. Once a specific gravity has been determined for a sample it is compared with the known SG for that species, to give a relatively accurate figure of how much cellulose was lost through bio-deterioration. The only complex part of this calculation is finding out what the wood species is in the waterlogged wood sample.

There are some simple ways to determine species of wood. First and easiest, is to look at the entire wooden artifact before it is sampled and determine from its gross anatomy what type of wood it is. Archaeologists and conservators can get a feel for this type of wood identification by picking up wood samples at a local lumber yard and studying and comparing the gross traits of each wood type. A more sophisticated method may include sending samples to various wood and forestry laboratories for microscopic analysis. If a microscope is available a conservation technician can prepare slides for microscopic comparison to the images presented in wood anatomy texts such as *The Textbook of Wood Technology*.

Specific Gravity

$$\frac{\text{Weight Oven Dry}}{\text{Waterlogged Volume}} = \text{SG}$$

In this formula it can be seen that specific gravity is a simple ratio of the oven-dried weight of a sample in grams divided by its waterlogged volume in cm^3. The waterlogged volume has not changed appreciably from the volume of the wood when it was first cut and worked.

When a specific gravity determination has been made and the species of wood sample is know cellulose content is an easy calculation. For example, if an artifact is determined to be white oak its SG should be close to .68 gr/cm^3 (there is some variation within species). If the archaeological white oak sample proves to have a calculated SG of .55 gr/cm^3 it is approximately 19% less dense than would be expected of un-degraded wood of that species. Bio-deterioration, therefore, has claimed about 19% of the complex carbohydrate structure in the wood. In other words it can be said that the sample contains 81% of its original cellulose. According to the flow chart on page 60 this archaeological wood is eligible for bulking.

3. Drying Behavior of a Sample (DB)

The behavior of a waterlogged wood sample on drying can be a good indicator of what to expect from the remainder of an archaeological artifact. In this test a sample of waterlogged wood is left to dry in the atmosphere. DB observations are both subjective and objective and sometimes require samples of larger dimensions than that typically used for the quantitative testing for moisture content and specific gravity.

Objective drying observations include measurements taken of the sample before and after drying with size reduction percentage tabulated. Pins can be placed in the wood to make the observations on size reduction more measurable and the grain of the wood should be carefully scrutinized so that observations include longitudinal, radial, and tangential size reduction measurements (see Wood Formation and Degradation—Theory). Size change above 15% indicates the wood has suffered damage on a microscopic level and the artifact should undergo treatment. But it is difficult to quantify all drying behavior, as it is different for each wood species. Many of the robust hardwoods for example, go through more drastic size reduction than softwoods that have the same percentage of remaining cellulose.

Drying samples should also be observed to note twisting, cracking and, face checking, all indicators that the wood has degraded. Face checking is a good indication that the surface of the artifact is in worse condition than the interior. A microscopic examination of the sample may reveal calcium carbonate concretion or corrosion residues on the bordered pits or elsewhere. Unfortunately a typical light microscope is not powerful enough to determine the condition of the bordered pits, but can show the viewer whether the cells in the wood have previously collapsed. Collapse is indicated if the normal uniform rectangular cell luminae have crumpled and become misshapen.

DB can be used to determine type of treatment for waterlogged wood, but there is far more guesswork involved than there is for the calculation of moisture content, specific gravity and cellulose percentage. The three waterlogged wood categories initially proposed by Christensen are included on the flow chart page 33 for this purpose.

BULKING TREATMENT—METHODOLOGY

Treatment for wood that remains in relatively sound condition (Christensen C, deJong class III.) is two fold. First, the partially degraded cell walls are to be strengthened. Second, the water remaining in the wood will be replaced with a substance that will lesson the capillary tension pressure exerted on the drying wood cell walls. Bulking accomplishes these goals. There are several accepted bulking agents in current use including polyethylene glycol (PEG) 300, 400, 600, 540 blend, and several varieties of sugar. These substances are all water-soluble and have molecular weights averaging less than 900. PEG numbers correspond roughly to their molecular weights, the exception being 540 blend. 540 blend is actually a mixture of PEG 400 and PEG 1450. Molecules that have a lower molecular weight than about 900 penetrate wood cell walls, whereas heavier weight molecules cannot.

The easiest to use bulking agents that have proven penetration ability for good condition waterlogged wood include liquid PEG 400 and granular sucrose (sugar-molecular weight 343). Of the two, sucrose or common white sugar is cheaper and easiest to find and gives the best results provided the treatment tank can be heated. At a minimum the treatment tank temperature should be maintained at about 43–50 degrees C or 110–120 F. At this treatment tank temperature biotic growth in the tank will remain minimal.

These agents are introduced into waterlogged wood with the same procedure. First a representative artifact is weighed (only one artifact can represent an entire batch). Weight gain is a ready indicator of bulking agent penetration. Artifacts are placed from storage into a treatment tank set up specifically for this purpose. The tank should have a means of heating and circulating (see Chapter 1, the Minimal Intervention Laboratory) and should be insulated and covered. Water and PEG or sucrose are placed in the tank at a 10% ratio, 5% for woods with light specific gravity's such as balsa, cork, or poplar. After the solution has been prepared a cover is put in place (bubble wrap makes a good insulated floating cover) and the stirring device is turned on.

For average to small artifacts of a few pounds or less than 1 kilogram the treatment tank percentage of bulking agent can be increased 10% per week until a 50% solution is achieved. Larger artifacts will require longer waiting times between solution increases, some up to a month. The reason for the gradual increase in bulking agent is to minimize osmotic pressure differentials between the fluids inside and outside the wood cells. Catastrophic results can be gained by immediately placing an artifact in too high a percentage bulking agent.

The representative artifact weighed before treatment should be weighed weekly after a 50% solution is achieved in the treatment tank. When artifact weight gain stabilizes at about 20% enough bulking agent has been absorbed for the artifact's dehydration in the humidity chamber. Yet some artifacts, those with the most missing cellulose, will show a greater weight gain of as much as 25%–35%.

NOTE ON SUCROSE BULKING

Sucrose has proven a cheap, reliable bulking agent. It does, however, have one drawback. At percentages less that 40% it tends to ferment. Careful observation of the treatment tank should be made each day. Runaway fermentation is easily recognized by foam production at the surface of the tank and a ruinous odor. Left untreated, fermentation will destroy the bulking capacity of the sugar and produce acetic acid, which is deleterious to the artifacts.

Heating the treatment tank to 50 degree C generally controls fermentation. Pasteurizing heat doses of up to 70 degrees C or 158 degree F for several hours are easily tolerated by most wooden artifacts. If the treatment tank is not capable of such temperatures 100 mls of Lysol per 50 liters of solution will eliminate fermentation for several weeks. As the sugar solution approaches a 40% concentration anti-biotic procedures are no longer necessary. The solution at that point becomes hypotonic to organisms and they can no longer grow and reproduce.

BULKING AND IMPREGNATION TREATMENT—METHODOLOGY

Waterlogged wood that contains half to three quarters of its original cellulose is invariably composed of stratified layers of wood that exhibit different amounts of deterioration. The outermost layer is the most degraded containing only a small amount of cellulose and can be quite soft. This layer is described as Christensen Class A or deJong class I. The next layer as you theoretically proceed into the wood is a thin zone of partially degraded wood in somewhat better condition than the outer layer but not as sound as the inner core. This wood is classified as Christensen class B or deJong class II. The inner core of an artifact in this category is likely sound wood in the same condition as wood treated in the far right column of the flow chart page 33. This is Christensen C or deJong class III.

Wood in a layered condition can be the most difficult to treat. Not only does the conservation treatment need to bulk the cell walls of the hard to penetrate inner most layer, but as little cellulose remains in the outer layer, it must be impregnated with a substance that will fill and harden all interstitial spaces plus the cell lumina.

Treatment on this type of wood can follow two paths depending on the thickness of the outer most degraded layer. If this layer is only a few millimeters in thickness the inner core is bulked following the same procedures and using the same bulking agents as outline in the treatment section preceding this section. However after the core of the artifact is bulked and before it is placed in the humidity chamber for dehydration a 50% PEG 3350 solution can be brush applied in order to consolidate and impregnate the outer most layer and prevent its cracking and face checking.

If the highly degraded zones run deeper, experience has demonstrated that treatment with PEG 540 blend is preferable. Since PEG 540 blend is a white paste mixture of PEG 400 and PEG 1450 it will penetrate and bulk the inner layer and impregnate the outer layers at the same time. Should cross checking on the surface become a problem, a brushed surface coat of PEG 3350 will be applied.

PEG 540 is introduced to the treatment tank in 10% increments working up to 50%, just as the bulking agents in the previous section were. Temperatures and time frame for adding agents are identical to those procedures outlined in the previous section.

As would be expected, however, treatment for this more degraded wood is not finished until a higher weight gain is achieved for the test artifacts. Treatment may take more time, depending on the size and permeability of the artifact, to both bulk and impregnate the artifact. Weight gains toward the upper end of the

25% to 40% spectrum are to be expected before stabilization. As in the bulking procedure the wood at this point is ready to be placed in the humidity chamber for slow dehydration.

IMPREGNATION TREATMENT—METHODOLOGY

Waterlogged wood in this category is entirely made up of the most highly degraded Christensen class A and deJong class I. Ironically, this is also among the easiest to conserve. The cell walls offer little impedance to the impregnating solution. In fact, there is so little cellulose in the secondary cell wall of the wood cells that it is impossible for them to remain rigid on dehydration. In this condition the only recourse left to the conservator concerned with maintaining the shape of the artifact is impregnation, the filling of all interstitial and cellular spaces with an inert substance that will harden on dehydration and support the porous wood structure.

Treatment follows the same procedures as those outlined for the previous two treatment sections. Impregnating agent will be initiated at 10% and worked up to 50% in 10% increments. In this case PEG 3350 is used. This heavier molecular weight PEG is too large a molecule to penetrate and bond with the wood cell wall but will do nicely to penetrate all of the empty spaces left in the degraded wood. PEG 3350 is a powder and must be placed in water to dissolve. Heating a peg 3350 solution will speed the dissolution of the powder.

As would be expected, an even higher artifact weight gain will signal the completion of the impregnation treatment. Weight gains of 30% to 50% will be experienced in the test artifacts. Unlike the previous two treatments, where the artifact is simply allowed to drip dry in the humidity chamber, excess PEG 3350 should be blotted from the surface of impregnated artifacts. This will keep the excess PEG from congealing.

HUMIDITY CHAMBER—METHODOLOGY

With the bulking and impregnating complete artifacts are removed from the treatment tank and allowed to drip into the tank. Great care should be taken in transporting the treated objects to the humidity chamber (HC—see Chapter 1). Delicate artifacts may need support or support trays of plexiglass or wood.

The artifacts will now be slow dried in the HC to lessen the possibility of osmotic collapse from drying too fast. Humidity chambers also allow artifacts to dehydrate at a uniform pace lessening stresses caused by the differing wood zones. The humidity chamber will contain shelving with enough space to accommodate the entire treatment tank. It is packed from the top down and is designed so that excess bulking and impregnating agent will drip off the artifacts through to the floor of the chamber for easy cleaning.

When the HC is occupied with artifacts from the treatment tank the humidifier is turned on. With the plastic curtain closed the humidifier will quickly bring the relative humidity up to near 100%. Each chamber should be tested before artifacts are introduced so that a dehydration rate, or that rate at which the chamber will gradually dry out, can be calculated. Ideally the chamber will gradually dry to ambient (about 50%) relative humidity in one month, or roughly 10% a week. Gages and control of the chamber's humidity are the responsibility of the conservator or conservation technician. Artifacts should be checked every few days and the internal atmosphere sprayed with Lysol to eliminate the possibility of mold growth. After four or five weeks in the gradually drying humidity chamber the artifacts are ready to be exposed to the atmosphere.

AIR DRY—METHODOLOGY

To initiate air drying the plastic curtain of the HC is pulled aside and the interior of the chamber exposed to the atmosphere in the laboratory. Ambient air in an air-conditioned building should remain at bout 50%. Anything lower than this will cause the wood to shrink, but should not cause any permanent damage. Artifacts can now be easily cleaned of excess bulking and impregnating agent. The higher weight PEGs can be melted and may need to be cleaned with a hand held blow dryer. After a week or two at ambient relative humidity the artifacts are prepared for storage in the artifact storeroom kept at 40%–50% relative humidity.

CONCLUSION

Wood can be a durable and lasting material that has seen use in artifacts throughout human history and pre-history. Dried wood can last for thousands of years with no appreciable loss of structure. Yet wood's organic nature lends itself to deterioration through biological processes, particularly when water is present. The complex carbohydrate cellulose that makes up the secondary wood cell walls is subject to attack by microorganisms. The wood can also become waterlogged in which case its deterioration is disguised until it dries. In waterlogged wood water takes the place of the missing cellulose, maintaining the artifact's original dimensions so long as it remains submerged. On drying the true nature of the waterlogged wood is revealed through destructive capillary tension collapse and subsequent dimensional changes.

The bulking and impregnating treatments described in this manual will offset the damage in most wood, controlling for collapse and shrinkage, at least to a large degree. As can be seen on page 34 there are nearly as many wood treatments as there are species of wood. This manual promotes sucrose and PEG as the simplest, least toxic, and most reversible of alternatives. The equipment needed for wood treatment at any particular laboratory depends on the condition of the wood that is recovered but should eventually include ways to treat both waterlogged and dry recovered material. As mentioned in Chapter 1, the Minimal Intervention Laboratory is always a work in progress, and in this case, the small expense needed to outfit a treatment tank and a humidity chamber will be more than offset by information gleaned from the artifacts saved.

CHAPTER 2: ARCHAEOLOGICAL WOOD

Aggebrandt, I.G. and O. Samuelson. "Penetration of Water-Soluable Polymers into Cellulose Fibers." *Journal of Applied Polymer Science* 8 (1964): 2801–2812.

Alagaria, Peitro. "The Conservation of the Treatment Tanks Used in the Conservation of the Wood of the Marsala Punic Ship." *Studies in Conservation* 22:3 (1977): 158–160. [ECU-182]

Albright, Alan B. "The Preservation of Small Waterlogged Wood Specimens with Polyethylene Glycol." *Curator* 9:3 (1966): 228–234. [ECU-11]

Ambrose, W.R. "The Treatment of Swamp Degraded Wood by Freeze Drying." (Report to the ICOM Committee for Conservation, Madrid) (1972).

—. "Stabilizing Degraded Swamp Wood by Freeze-Drying." *ICOM Proceedings* (4th Triennial Meeting, Venice) (1975): 1–14.

—. "Freeze Drying of Swamp Degraded Wood." *Conservation of Wood and Stone* (IIC, London) (1975): 53–57.

—. "Sublimation Drying of Degraded Wet Wood." *Proceedings* (Pacific Northwest Conference 1) (1976): 7–15. [ECU-239]

Amoigon, J. and P. Larrat. "The Treatment of Waterlogged Wood by Lyophilization Under Normal Atmospheric Pressure—Application to Large Size Objects." *ICOM-WWWG Proceedings* (Grenoble) (1984): 181–186.

Anon. "Two PEG's are Better than One," *Chemistry in Britain*, (1987). [ECU-107]

Anon. *Carbowax Polyethylene Glycols*. Dunbury, Connecticut: Union Carbide Corporation, Ethylene Oxide Derivitives, (1981).

Arrhenius, O. "Corrosion on the Warship *Wasa*." *Bulletin* (Swedish Corrosion Institute 48) (1967).

Barbour, James R. "The Condition and Dimensional Stabilization of Highly Deteriorated Waterlogged Hardwoods." *ICOM-WWWG Proceedings* (Grenoble) (1984): 23–37. [ECU-13]

—. "Treatments for Waterlogged and Dry Archaeological Wood." *Archaeological Wood* (1990): 177–192. [ECU-290]

Barbour, James R. and L. Leney. "Shrinkage and Collapse in Waterlogged Archaeological Wood: Contribution III Hoko River Series." *ICOM-WWWG Proceedings* (Ottawa) (1981): 209–225. [ECU-14]

Barclay, R., R. James, and A. Todd. "The Care of Wooden Objects." *CCI Bulletin* 8 (1980): 1–15. [ECU-15]

Barka, G., H. Knauer, and P. Hoffman. "Simultaneous Separation of Oligometric and Polymeric Ethylene Glycols (DP 1–110) Using Reverse Phase HPLC (HRPLC)." *Journal of Chromatography* (1987).

Barker, Harold. "Early Work on the Conservation of Waterlogged Wood in the U.K." *MMR* (National Maritime Museum, Greenwich, U.K.) 16 (1975): 61–63. [ECU-252]

Barkman, L. "Preservation of the Warship *Wasa*." *Maritime Monographs and Reports* (National Maritime Museum, Greenwich, U.K.) 16 (1975): 65–106.

Barkman, L., S. Bengtsson, B. Hafors, and B. Lundvall. "Processing of Waterlogged Wood." *Proceedings* (Pacific Northwest Conference 1) (1976): 17–26. [ECU-240]

Bateson, B.A. "Sorbitol in Wood Treatment." *Chemical Trade Journal and Chemical Engineer* (1938): 26–27. [ECU-406]

—. "The Shrinkage and Swelling of Wood—Experiments on the Influence of Sorbitol." *CRJCE* [n.d.].

Baynes-Cape, A.D. "Fungicides and the Preservation of Waterlogged Wood." *Maritime Monographs and Reports* (National Maritime Museum, Greenwich, U.K.) 16 (1975): 31–33. (ECU-256]

Bengtsson, Sven. "What About *Wasa*?" *Realia* 1 (1990): 309–311.

Berry, I.S. and P.R. Houghton, eds. "Conservation of the Timbers of the Tudor Ship *Rose*." *Papers* (Biodeterioration 6: Presented at the 6th International Biodeterioration Symposium, Slough, U.K.) (1986): 354–362.

Biek, L. "Some Notes on the Freeze-Drying of Large Timbers." *Maritime Monographs and Reports* (National Maritime Museum, Greenwich, U.K.) 16 (1975): 25–29. [ECU-257]

Björdal, C.G. and T. Nilsson. "Observations on Microbial Growth During Conservation Treatment of Waterlogged Archaeological Wood." *Studies in Conservation*. Vol. 46 No. 3, 2001. [ECU-632]

Blackshaw, S.M. "Comparison of Different Makes of PEG and Results on Corrosion Testing." *MMR* (National Maritime Museum, Greenwich, U.K.) 16 (1975): 51–58.

—. "Coments on the Examination and Treatment of Waterlogged Wood Based on Work Carried Out During the Period 1972–1976 at the British Museum." *Proceedings* (Pacific Northwest Conference 1) (1976): 27–34. [ECU-0238]

Blanchette, Robert, John E. Haight, Robert J. Koestler, Pamela B. Hatchfield, Dorthea Arnold. "Assessment of Deterioration in Archaeological Wood from Ancient Egypt." *Journal of the American Institute for Conservation*. Vol. 33 No. 1, 1994. [ECU-633]

Bolton, A.J. and J.A. Petty. "A Model Describing Axial Flow of Liquids Through Conifer Wood." *Wood Science and Technology* 12 (1978): 37–48. [ECU-212]

"Borate Wood Preservatives—Marine Applications?" *Wooden Boat* Vol. 110, 1993, pp. 92–93. [ECU-767]

Borgin, K. "Mombasa Wreck Excavation: Second Preliminary Report." *International Journal of Nautical Archaeology* 7 (1978): 301–319.

Bourgiba, N. and P.A. Hodges. "The Penetration of Fluids with Different Swelling Characteristics into Timber." *Journal of the Institute of Wood Science*. Vol. 15 No. 3, 2000. [ECU-634]

Bright, Leslie S. "Recovery and Preservation of a Fresh Water Canoe." *International Journal of Nautical Archaeology* 8 (1979): 47–57.

Brownstein, Allen. "The Chemistry of Polyethylene Glycol." *ICOM-WWWG* (Proceedings, Ottawa) (1981): 279–285. [ECU-18]

Browse, David S. "Archaic Dugout Canoe Found in Northern Ohio." *The Explorer* 20:2 (1978): 13–17.

Bryce, T., and H. McKerrell. "The Acetone-Rosin Method for the Conservation of Waterlogged Wood and Some Thoughts on the Penetration of PEG into Oak." *Maritime Monographs and Reports* (National Maritime Museum, Greenwich, U.K.) 16 (1975): 35–43. [ECU-255]

Carbowax Polyethylene Glycols. Dunbury, Connecticut: Union Carbide Corporation, Ethylene Oxide Derivatives, 1981.

Caple, Chris and Will Murray. "Characterization of a Waterlogged Charred Wood and Development of a Conservation Treatment." *Studies in Conservation*. Vol. 39 No. 1, 1994. [ECU-635]

CCI Laboratory Staff. "Making Padded Blocks." *CCI Notes* (Canadian Conservation Institute) 11/2 (1995) [ECU-540]

"Chemical Wood Deterioration." *Wooden Boat* Vol. 154, 2000, pp. 114–115. [ECU-768]

Chen, Peter Y.S. and Yifu Tang. "Variation in Longitudinal Permeability of Three U.S. Hardwoods." *Forest Products Journal* 41:11/12 (1991). [ECU-438]

Chen, Zhanjing, Eugene M. Wengert, and Fred M. Lamb. "A Technique to Electrically Measure the Moisture Content of Wood Above Fiber Saturation." *Forest Products Journal*. Vol. 44 No. 9, 1994. [ECU-636]

Choong, B.T. "Effect of Extractives on Shrinkage and Other Hygroscopic Properties on Ten Southern Pine Woods." *Wood and Fiber* 1:2 (1970).

Choong, E.T., J.F.G. Mackay, and C.M. Stewart. "Collapse and Moisture Flow in Kiln-Drying and Freeze-Drying of Woods." *Wood Science* 6:2 (1973): 127–135. [ECU-497]

Christensen, B. Brorson. *Conservation of Waterlogged Wood at the National Museum of Denmark*. Copenhagen: National Museum of Denmark, 1970.

—. "Developments in the Treatment of Waterlogged Wood in the National Museum of Denmark, 1962–1969." *Proceedings* (IIC Conference on Stone and Wooden Objects, New York, 1970, 2nd Edition, IIC, London) (1971): 29–35.

Clarke, R.W. and C. Gregson. "A Large Scale PEG Conservation Facility for Waterlogged Wood at the National Maritime Museum, Greenwich." *ICOM Proceedings* (8th Triennial Meeting) (1987): 301–307.

Clarke, R.W. and J.P. Squirrell. "A Theoretical and Comparative Study of Conservation Methods for Large Waterlogged Wooden objects." *ICOM-WWWG Proceedings* (Ottawa) (1981): 19–27. [ECU-19, 579]

—. "The Pilodyn—An Instrument for Assessing the Condition of Waterlogged Wooden Objects." *Studies in Conservation* 30:4 (1985): 177–183. [ECU-310]

Colardelle, M. "The Center of Study and the Treatment of Waterlogged Wood." *ICOM-WWWG Proceeding* (Grenoble) (1984): 17–19.

Coles, John M. "The Somerset Levels: A Waterlogged Landscape." *ICOM-WWWG Proceedings* (1981): 129–141. [ECU-21]

—. "Session III. Conservation for Frozen or Wetland Archaeological Sites: Introductory Talk." *ICOM-WWWG Proceedings* (Ottawa) (1981): 121. [ECU-20]

Clydesdale, A. "A New Technique for Wood Conservation." *Scottish Society for Conservation and Restoration* Vol. 6 No. 1, 1995. [ECU-769]

Comstock, Gilbert L. "Physical and Structural Aspects of the Longitudinal Permeability of Wood." (PhD Thesis, College of Environmental Science and Forestry, Syracuse University) (1968).

Comstock, Gilbert L. and W.A. Cote. "Factors Affecting Permeability and Pit Aspiration in Coniferous Sap-Wood." *Wood Science and Technology* 2 (1968): 219–239. [ECU-215]

Cook, C. and David Grattan. "A Practical Comparitive Study of Treatments for Waterlogged Wood, Part III: Pretreatment Solutions for Freeze-Drying." *ICOM-WWWG Proceedings* (Grenoble) (1984): 219–239. [ECU-23]

Cott, J. "The Conservation of Wet Wood." *Poseidon* 77 (1968): 310–311.

Cutler, D.F. "The Anatomy of Wood and the Processes of its Decay." *Maritime Monographs and Reports* (National Maritime Museum, Greenwich, U.K.) 16 (1975): 1–7. [ECU-259]

Daley, T.W. and Lorne D. Murdock. "Underwater Molding and an X-Section of the *San Juan* Hull: Red Bay, Labrador, Canada." *ICOM Proceedings* (Copenhagen) (1984).

Daley, T., Murdock, L.D. and C. Newton. "Underwater Molding Techniques on Waterlogged Ships Timbers Employing Various Products Including Liquid Polysulfide Rubber." *ICOM-WWWG Proceedings* (Ottawa) (1981): 39–40. [ECU-25]

Daugherty, R.D. and D.R. Croes. "Wet Sites in the Pacific Northwest." *Proceedings* (Pacific Northwest Conference 2) (1976): 17–29. [ECU-227]

Davidson, R.L., ed. "Section 19–1: Polyethylene Glycol." *Handbook of Water Soluable Gums and Resins*. New York: McGraw-Hill, 1980.

Dawson, J.E. "Some Considerations in Choosing a Biocide." *ICOM-WWWG Proceedings* (Ottawa) (1981): 269–277. [ECU-28]

Dawson, J.E., R. Ravindra, and R.H. LaFontaine. "A Review of Storage Methods for Waterlogged Wood." *ICOM-WWWG Proceedings* (Ottawa) (1981): 227–235. [ECU-27]

Denton, Mark H. and Joan S. Gardner. "The Recovery and Conservation of Waterlogged Goods for the Well Excavated at the Fort Loudoun Site." *Historical Archaeology* 17:1 (1983): 96–103. [ECU-382]

Dersarkissian, M. and Mayda Goodberry. "Experiments in Non-Toxic Anti-Fungal Agents." *SIC* 25 (1980): 28–36. [ECU-174]

Dorge, Valerie, and F. Carey Howlett. "Painted Wood: History and Conservation." *WAG AIC Symposium Williamsburg*, 1994. [ECU-770]

Eaton, John W. "The Preservation of Wood by the Alum Process." *Florida Anthropologist* 15:4 (1962): 115–117. [ECU-0029]

Eenkhoom, W. and A.J.M. Wevers. "Report of an Investigation into Wood and Wood Samples in the Area of the East Indiaman *Amsterdam*." (Rijksdienst voor Ijsselmerpolders Report, No. 28) (1983).

Eenkhoom, W., J. De Jong, and A.J.M. Wevers. "The Wood of the VOC Ship *Amsterdam*." *Conservation of Waterlogged Wood* (1979): 85–90.

Eilwood, E.L., B.A. Ecklund, and E. Tavarin. "Collapse in Wood Exploratory Experiments in its Prevention." *Forest Products Journal* 10:1 (1960): 8–21.

Erickson, R.W., R.L. Hossfeld, and R.M. Anthony. "Pretreatment with Ultrasonic Energy—Its Effect Upon Volumetric Shrinkage of Redwood." *Wood and Fiber* 2:1 (1970): 12–18.

Fenwick, V., ed. "The Graveny Boat." *British Archaeological Reports* (National Maritime Museum, Greenwich, U.K., Archaeological Report Series No. 3, British Series 53) (1978).

Finney, R.W. and A.M. Jones. "Direct Analysis of Wood Preservatives in Ancient Oak from the *Mary Rose* by Laser Microprobe Mass Spectrometry." *Studies in Conservation*. Vol. 38 No. 1, 1993. [ECU-637]

Florian, Mary-Lou E., et al. "The Physical, Chemical, and Morphological Condition of Marine Archaeological Wood Should Dictate the Conservation Process." *Papers* (First Southern Hemisphere Conference on Maritime Archaeology) (1978): 128–144.

Florian, Mary-Lou E., and R. Renshaw-Beauchamp. "Anomalous Wood Structure: A Reason for Failure of PEG in Freeze-Drying Treatments of Some Waterlogged Wood from the Ozette Site." *ICOM-WWWG Proceedings* (Ottawa) (1981): 85–98.

Fogarty, W.M. "Bacteria, Enzymes, and Wood Permeability." *Process Biochemistry* (1973): 30–34. [ECU-403]

Fox, Louise. "The Acetone-Rosin Method for Treating Waterlogged Hardwoods at the Historic Resource Convention Branch, Ottawa." *ICOM Proceedings* (Fremantle) (1987): 73–94. [ECU-0198]

Friedman, Janet P. "Wood Identification—An Introduction," *Proceedings* (Pacific Northwest Conference 1) (1976): 67–79. [ECU-234]

Furuno, T. and T. Goto. "Structure of the Interface Between Wood and Synthetic Polymers: V. The Penetration of Polyethylene Glycol into Woody Cell Wall." *Journal of the Japan Wood Research Society* 20:9 (1974): 446–452. [ECU-401]

"The Galilee Boat: A Painstaking Restoration." *Archaeology* 42:1 (1989): 18. [ECU-347]

General Discussion Period. "The Conservation of Shipwrecks." *ICOM-WWWG Proceedings* (Ottawa) (1981): 61–66. [ECU-102]

—. "Conservation for Frozen or Wet Land Archaeological Sites." *ICOM-WWWG Proceedings* (Ottawa) (1981): 181–188. [ECU-104]

—. "Progress in the Treatment of Waterlogged Wood: Biocides." *ICOM-WWWG Proceedings* (Ottawa) (1981): 267–268. [ECU-105]

General Discussion Period, Session II. "Analysis and Classification of Wood." *ICOM-WWWG Proceedings* (Ottawa) (1981): 117–120. [ECU-103]

Gerrasimova, N.G. "On the Investigation of Waterlogged Wood in Connection with its Conservation." *ICOM Proceedings* (7th Triennial Meeting, Copenhagen) (1984).

Gerrasimova, N.G., E.A. Mikolajchuk, and M.I. Kolosova. "On the Conservation of Archaeological Wood by Introduction of Waxlike Substances Into It." *ICOM Proceedings* (6th Triennial Meeting, Ottawa) (1981): 1–10.

Gilroy, D. "Conservation of a Pulley Sheave from the Dutch East Indiaman *Zeewiyk*, 1727." *ICOM Bulletin* 4 (12/1978): 25–27.

Ginier-Gillet, A., M.D. Parchas, R. Ramiere, and Q.K. Tran. "Conservation Methods Developed at the C.T.B.G.E. (Grenoble, France), Impregnation of Radiocarbon Resin and Lyphilization." *ICOM-WWWG Proceedings* (Grenoble) (1984): 125–139.

Gowers, H.J. "Problems Concerning the Conservation of Wood." (IIC-UK Group Bulletin Offprint) (1974): 2p.

Grattan, David W. "A Practical Comparative Study of Treatments for Waterlogged Wood, Part II: The Effect of Humidity on Treated Wood." *ICOM-WWWG Proceedings* (Ottawa) (1981): 243–252. [ECU-35]

—. "A Practical Comparative Study of Several Treatments for Waterlogged Wood." *Studies in Conservation* 27:3 (1982): 124–136. [ECU-333]

—. "The Degradation of Waterlogged Wood, Recent Progress in Conserving Waterlogged Wood." *Museum* 1:137 (1983): 22–26. [ECU-400]

—. "The ICOM Waterlogged Wood Working Group." *Proceedings* (13th Conference on Underwater Archaeology, Society of Historical Archaeology) (1984): 8–10. [ECU-320]

—. "Advances in the Conservation of Waterlogged Wood, 1981–1984." *ICOM Proceedings* (7th Triennial Meeting, Copenhagen) (1984).

—. "Treatment of Waterlogged Wood." *Proceedings* (First International Conference of Wet Site Archaeology, University of Florida, Gainesville) (1986).

—. "Some Observations on the Conservation of Wooden Shipwrecks." *ICOM Bulletin* 12 (1986).

—. "Waterlogged Wood." *Conservation of Marine Archaeological Objects* C. Pearson, ed. (London: Butterworths Series in Conservation and Museology) (1987): 55–67. [ECU-33]

—. "International Comparative Wood Treatment Study." *ICOM Proceedings* (Fremantle) (1987): 153–203. [ECU-204]

—. "Moisture Measurement Using Embedded Moisture Probes." *CCI Newsletter* (1989): 11–12. [ECU-32]

Grattan, David W., ed. *Proceedings* (ICOM Waterlogged Wood Working Group Conference. Ottawa: International Council of Museums, Comittee for Conservation, Working Group on Waterlogged Wood) (1981). [ECU-2]

Grattan, David W. and R.W. Clarke. "Conservation of Waterlogged Wood." *Conservation of Marine Archaeological Objects* C. Pearson, ed. (London: Butterworths Series in Conservation and Museology) (1987): 164–206. [ECU-36]

Grattan, David W. and S. Drouin. "Conserving Waterlogged Wood—30 Million Years Old." *ICOM Proceedings* (Fremantle) (1987): 61–72. [ECU-199]

Grattan, David W. and C. Mathias. "Analysis of Waterlogged Wood: The Value of Chemical Analysis and Other Simple Methods in Evaluating Condition." *Somerset Levels Papers* 12 (1986): 6–12; 106–108. [ECU-37]

Grattan, David W. and J.C. McCawley. "The Potential of the Canadian Winter Climate for the Freeze-Drying of Degraded Waterlogged Wood." *Studies in Conservation* 23 (1978): 157–167. [ECU-178]

Grattan, David W., J.C. McCawley, and C. Cook. "The Potential of the Canadian Winter Climate for the Freeze Drying of Degraded Waterlogged Wood: Part II." *Studies in Conservation* 25:3 (1980): 118–136. [ECU-337]

—. "The Conservation of a Waterlogged Dug-out Canoe Using Natural Freeze-Drying." *ICOM-WWWG Proceedings* (Ottawa) (1981).

Greaves, H. "A Review of the Influence of Structural Anatomy on Liquid Penetration into Hardwoods." *Journal of the Institute of Wood Science* 6:6 (1974).

Gregson, C.W. "Progress on the Conservation of the Graveney Boat." *Maritime Monographs and Reports* (National Maritime Museum, Greenwich, U.K.) 16 (1975): 113. [ECU-248]

Grimstad, Kirsten, ed. "A Large Scale Polyethylene Glycol Conservation Facility for Waterlogged Wood at the National Maritime Museum, Greenwich, U.K.." *ICOM Proceedings* (8th Triennial Meeting, Sydney) (1987).

Grosso, G.H. "Volume Processing of Waterlogged Wood at a Remote Archaelogical Site: Modification of Old Techniques, Identification of Special Problems, and Hopes for their Solution." *Proceedings* (Pacific Northwest Conference 1) (1976): 35–48. [ECU-237]

—. Discussion Session: "Field Conservation at Ozette." *Proceedings* (Pacific Northwest Conference 2) (1976): 91–97. [ECU-230]

—. "Experiments with Sugar in Conserving Waterlogged Wood." *ICOM-WWWG Proceedings* (Ottawa) (1981): 1–9.

Grosso, G.H., ed. *Proceedings* (Pacific Northwest Wet Site Wood Conservation Conference, Neah Bay, Washington, Vol. 1 & 2) (1976). [ECU-262 and 263]

Haas, A. and H. Muller-Beck. "A Method for Wood Preservation Using Arigal-C." *Studies in Conservation* 5 (1960): 150–158. [ECU-409]

Hafors, Birgitta. "The Drying of the Outer Planking of the *Wasa* Hull." *ICOM Proceedings* (Grenoble) (1984): 313–321. [ECU-38]

—. "The Role of the *Wasa* in the Development of the Polyethylene Glycol Preservation Method." *Archaeological Wood (*1990): 195–216. [ECU-294]

Hamilton, Donny L. *Basic Methods of Conserving Underwater Archaeological Material Culture.* Wahsington, D.C.: U.S. Department of Defense Legacy Resource Management Program, January 1996. [ECU-610]

Hart, C.A. "Relative Humidity, EMC, and Collapse Shrinkage in Wood." *Forest Products Journal* 34:11/12 (1985): 45–54. [ECU-399]

Harvey, P. "A Review of Stabilisation Works on the Wreck of the *William Salthoue* in Port Phillip Bay." *Bulletin of the Australian Institute for Maritime Archaeology* Vol. 20, 1996. [ECU-771]

Hawley, Greg. "The Recovery, Stabilization, and Preservatrion of Large Wooden Structure and Cargo from the Steamboat *Arabia.*" (Unpublished Manuscript, *Arabian* Steamboat Museum, Kansas City, Missouri) [n.d.]. [ECU-517]

Hawley, L.F. "Wood-Liquid Relations." (USDA Technical Bulletin No. 248, Washington D.C.) (1931).

Hicken, Norman E. *Wood Preservation: A Guide to the Meaning of Terms.* London: Hutchison, 1971.

—. *The Woodworm Problem.* London: Hutchison (1972).

Hillman, D. and Mary-Lou E. Florian. "A Simple Conservation Treatment for Wet Archaeological Wood." *Studies in Conservation* 30 (1985): 39–41. [ECU-405]

Hoffman, Per. "Short Note on the Conservation Program for the Bremen Cog." *Conservation of Waterlogged Wood* (1979): 41–44. [ECU-299]

—. "Chemical Wood Analysis as a Means of Characterizing Archaeological Wood." *ICOM-WWWG Proceedings* (Ottawa) (1981): 69–72. [ECU-40]

—. "A Rapid Method for the Detection of Polyethylene Glycols (PEG) in Wood." *Studies in Conservation* 18:4 (1983): 189–193. [ECU-122]

—. "On the Stabilization of Waterlogged Oakwood with PEG:Molecular Size vs. Degree of Degradation." *ICOM Proceedings* (Grenoble) (1984): 95–115. [ECU-40]

—. "On the Stabilization of Waterlogged Oakwood with PEG; II: Designing a Two-Step Treatment for Multi-Quality Timbers." *Studies in Conservation* 31:3 (1986): 103–113. [ECU-189]

—. "HPLC for the Analysis of Polyethylene Glycols (PBG) in Wood." *ICOM Proceedings* (Fremantle) (1987): 41–60. [ECU-197]

—. "To be and to continue being a cog: the conservation of the Breman Cog of 1318." *International Journal of Nautical Archaeology.* Vol. 30 No. 1, 2001, pp. 129–140. [ECU-638]

Hoffman, P. and M.A. Jones. "Structure and Degradation Process for Waterlogged Archaeological Wood." *Archaeological Wood* (1990): 99–104. [ECU-291]

Hoffman, Per, Kwang-nam Choi, and Yong-han Kim. "The 14th-century Shinan Ship—Progress in Conservation." *International Journal of Nautical Archaeology* 20:1 (1991): 59–64. [ECU-491]

Hoffman, Per and Robert A. Blanchette. "The Conservation of a Fossil Tree Trunk." *Studies in Conservation.* Vol. 42 No.2, 1997, pp. 74–82. [ECU-639]

Hoyle, Robert J. "Relating Wood Science and Technology to the Conservator." *Proceedings* (Pacific Northwest Conference 2) (1976): 99–104. [ECU-231]

Hug, B. "Lyphilization: 10 Years of Experience." *ICOM-WWWG Proceedings* (Grenoble) (1984): 207–212.

ICOM Comittee for Conservation Working Group on Wet Organic Archaeological Materials. *Newsletter* No. 1–20 (1978–1990). [ECU-297]

Inglis, R., Croes, D., and C. Rose. "Round Table Discussion." *Proceedings* (Pacific Northwest Conference 2) (1976): 81–90. [ECU-229]

Conservation of Waterlogged Wood (International Symposium on the Conservation of Large Objects of Waterlogged Wood, Amsterdam, September 1979. The Hague: Government Printing Conference) (1981).

Inverarity, R.B. "The Conservation of Wood from Fresh Water." *Diving Into the Past*. St. Paul: Minnesota Historical Society, 1964.

Irwin, H.T. and G. Wessen. "A New Method for the Preservation of Waterlogged Archaeological Remains: Use of Tetraethyl Orthosilicate." *Proceedings* (Pacific Northwest Conference 1) (1976): 49–59. [ECU-0236]

Ishimaru, Y. "Adsorption of Polyethylene Glycol on Swollen Wood: I. Molecular Weight Dependence." *Journal of the Japan Wood Research Society* 22:1 (1976): 22–28. [ECU-388]

Jagels, Richard. "A Deterioration Evaluation Procedure for Waterlogged Wood." *ICOM-WWWG Proceedings* (Ottawa) (1981): 69–72. [ECU-43]

Jamiere, G. and G. Meurges. "The Treatment of Waterlogged Wood by Freeze-Drying at the Natural History Museum." *ICOM-WWWG Proceedings* (Grenoble) (1984): 175–184.

Jenssen, Victoria. "Giving Waterlogged Wood the Brush." *ICOM-WWWG Proceedings* (Ottawa) (1981).

Jenssen, Victoria, and Lorne Murdock. "Review of the Conservation of *Machault*'s Ships Timbers, 1973–1981." *ICOM-WWWG Proceedings* (Ottawa) (1981): 41–49. [ECU-44]

Jespersen, Kirsten. "Conservation of Waterlogged Wood by Use of Tertiary Butanol, PEG, and Freeze-Drying." *Conservation of Waterlogged Wood* (1979): 69–76. [ECU-312]

—. "Some Problems of Using Tetraethozysilane (Tetra Ethyl Ortho Silicate: TEOS) for Conservation of Waterlogged Wood." *ICOM-WWWG Proceedings* (Ottawa) (1981): 203–207. [ECU-0046]

—. "Extended Storage of Waterlogged Wood in Nature." *ICOM Proceedings* (Grenoble) (1984): 39–54. [ECU-45]

—. "Precipitation of Iron-Corrosion Products on PEG-Treated Wood." *ICOM Proceedings* (Fremantle) (1987): 141–152. [ECU-202]

Johnson, Bruce R., Rebecca E. Ibach, and Andrew J. Baker. "Effect of Salt Water Evaporation on Tracheid Separation from Wood Surfaces." *Forest Products Journal* 42:7/8 (1992). [ECU-437]

Johnson, Rosemarie. "Large Waterlogged Timber Structures Excavated on the Billingsgate Excavation, London, 1982." *ICOM Proceedings* (Grenoble) (1984): 63–69. [ECU-47]

Jones, A.M. and E.B.G. Jones. "Conservation of Timbers of the Tudor Ship *Mary Rose*." *Proceedings* (Sixth International Biodeterioration Symposium, Program and Abstracts) Vol. 6 Abstract No. 60 (1984).

Jones, E.B.G. "The Decay of Timber in Aquatic Environments." *Report* (British Wood Preserving Association Annual Convention) (1972): 31–49.

de Jong, J. "The Conservation of Waterlogged Timber at Ketelhaven (Holland)." *ICOM Proceedings* (4th Triennial Meeting, Venice) (1975): 1–9.

—. "Conservation Techniques for Old Waterlogged Wood from Shipwrecks Found in the Netherlands." *Biodeterioration Inves. Tech.* (1977): 295–338.

—. "The Deterioration of Waterlogged Wood and its Protection in the Soil." *Conservation of Waterlogged Wood* (1979): 31–40. [ECU-300]

—. "The Conservation of Shipwrecks by Impregnation with Polyethylene Glycol." *Conservation of Waterlogged Wood* (1979): 57–67. [ECU-317]

Jover, Anna. "The Application of PEG 4000 for the Preservation of Paleolithic Wooden Artifacts." *Studies in Conservation*. Vol. 39 No. 3, 1994. [ECU-640]

Kahanov, Yaacov. "Wood Conservation of the Ma'agan Mikhael Shipwreck." *International Journal of Nautical Archaeology*. Vol. 26 No. 4, 1997, pp. 316–329. [ECU-641]

Katzev, M.L. "Resurrecting the Oldest Known Greek Ship." *National Geographic* 137:6 (1970): 841–847.

Kazanskaya, S.Y. and Klara F. Nikitina. "On the Conservation of Waterlogged Degraded Wood by a Method Worked out in Minsk." *ICOM-WWWG Proceedings* (Grenoble) (1984): 139–146. [ECU-48]

Kawai, T. and F. Masuzawa. "Study on Conservation of Waterlogged Wood with Freeze Drying at Reduced Pressure." *Conservation Science Bulletin* 3 (1974): 59–67.

Kaye, Barry and David J. Cole-Hamilton. "Conservation of Knife Handles from the Elizabethan Warship *Makeshift*." *International Journal of Nautical Archaeology*. Vol. 24 No. 2, 1995, pp. 147–158. [ECU-642]

Kaye, Barry, David J. Cole-Hamilton, and Kathryn Morphet. "Supercritical Drying: A New Method for Conserving Waterlogged Archaeological Materials." *Studies in Conservation*. Vol. 45 No. 4, 2000, pp 233–252. [ECU-643]

Keene, Susanne. "An Approach to the Sampling and Storage of Waterlogged Timbers from Excavation." *The Conservator* 1 (1977): 8–11. [ECU-410, 577]

—."Waterlogged Wood From the City of London." *ICOM-WWWG Proceedings* (Ottawa) (1981): 177–188. [ECU-49]

Kelly, John. "The Construction of a Low Cost, High Capacity Vacuum Freeze Drying System." *Studies in Conservation* 25 (1980): 176–179. [ECU-125]

Kotlik, Petr. "Impregnation Under Low Pressure." *Studies in Conservation*. Vol. 43 No. 1, 1998, pp. 42–47. [ECU-644]

Kubler, Hans. "Corrosion of Nails in Wood Construction Interfaces." *Forest Products Journal*. Vol. 42 No. 1, 1992, pp. 47–49. [ECU-645]

Kucra, L.J. "Moisture Conditions in PEG-Treated Wood from the Warship *Wasa*." *Xylorama Trends in Wood Research* (Stockholm: Royal Institute of Technology) (1985): 192–197.

de La Baume, Sylvia. "Archaeological Wood Desalting by Electrophoresis." *ICOM Proceedings* (Fremantle) (1987): 153–162. [ECU-203]

Lan, Zhang. "A Note on the Conservation of a Thousand Year Old Boat." *Studies in Conservation*. Vol 40 No. 3, 1995, pp. 201–206. [ECU-646]

Larsen, Knut Einar. *Conservation of Historic Timber Structures, an Ecological Approach*. London: Elsevier, 2000.

Laures, F.F. "A Means of Rapid Drying of Antique Wooden Material Previously Saturated with Polyethylene Glycol (PEG 4000) in a Watery Solution." *International Journal of Nautical Archaeology* 13:4 (1984): 325–327. [ECU-330]

—. "More Details about Rapid Drying of Wood by "Frying" it in Polyethylene Glycol." *International Journal of Nautical Archaeology* 15:1 (1986): 68–69. [ECU-155]

Lodewijks, J. "Ethics and Aesthetics in Relation to the Conservation and Restoration of Waterlogged Wooden Shipwrecks." *Conservation of Waterlogged Wood* (1979): 107–110. [ECU-313]

Lous, F. "The Recovery of the *Amsterdam*." *Conservation of Waterlogged Wood* (1979): 91–97. [ECU-316]

Lucas, Deirdre A. "On Site Packing and Protection for Wet and Waterlogged Wood." *ICOM-WWWG Proceedings* (Ottawa) (1981): 51–55. [ECU-53]

MacDonald, George F. "The Management of Wet Site Archaeological Resources." *ICOM-WWWG Proceedings* (Ottawa) (1981): 123–128. [ECU-54]

MacLeod, Ian Donald. "Hygroscopicity of Archaeological Timber: Effects of Molecular Weight of Impregnant and Degree of Degradation." *ICOM Proceedings* (Fremantle) (1987): 211–213.

—. "Conservation of Waterlogged Timbers from the *Batavia*, 1629." *Bulletin of the Australian Institute for Maritime Archaeology* 14:2 [n.d.]: 1–8.

MacLeod, Ian Donald, ed. *Proceedings* (ICOM Waterlogged Wood Working Group Conference. Fremantle: Western Australia Museum) (1987). [ECU-264]

MacLeod, Ian Donald and D.R. Gilroy. "Colour Measurement of Treated and Air-Dried Wood." *ICOM Proceedings* (Fremantle) (1987): 203–213. [ECU-205]

MacLeod, Ian D., Fiona M. Fraser, and Vicki L. Richards. "The PEG-Water Solvent System: Effects of Composition on Extraction of Chloride and Iron from Wood and Concretion." *ICOM Proceedings* (Fremantle) (1987): 245–263.

Mahrer, N. "The Conservation of Recently Discovered Shipwreck Material from Jersey." *Scottish Society for Conservation and Restoration* Vol. 11 No. 4, 2000. [ECU-772]

Marsden, Peter. "Post-Medieval Ship Archaeology: The Present and the Future." *Conservation of Waterlogged Wood* (1979): 25–29. [ECU-301]

Masuzawa, F. and Y. Nishiyama. "Experiments on the Impregnation of Waterlogged Wood with PEG (II)." *Conservation Science Bulletin* 3 (1974): 39–46.

Matsuda, Takatsuga. "On Reproducing the Shape of the Tissues of the Dried and Shrunk Waterlogged Wood for the Identification of its Species." *ICOM Proceedings* (Grenoble) (1984): 55–62. [ECU-55]

Mavroyamakis, E.G. "Aging of Reinforced Ancient Waterlogged Wood by Ray Methods." *ICOM-WWWG Proceedings* (Ottawa) (1981): 263–266. [ECU-56]

McCawley, J. Cliff. "Waterlogged Artifacts: The Challenge to Conservation." *JCCI* 2 (1977): 17–36. McCawley, J.C., and David W. Grattan. "Natural Freeze-Drying: Saving Time, Money, and a Waterlogged Canoe." *Journal of the Canadian Conservation Institute* 4 (1980): 36–39.

McCawley, J. C., David W. Grattan, and Clifford Cook. "Some Experiments in Freeze-Drying: Design and Testing of a Non-Vacuum Freeze Dryer." *ICOM-WWWG Proceedings* (Ottawa) (1981):253–262. [ECU-57]

McClelland, C.P. and R.L. Bateman. "Technology of the PEGs and Carbowax Compounds." *Chemical and Engineerina News* 23:3 (1945): 247–251.

McGrail, Sean. "Progress Towards a Centre for the Conservation of Waterlogged Wood." *MMR* (National Maritime Museum, Greenwich, U.K.) 16 (1975): 107–108. [ECU-250]

McGrail, Sean and C. Gregson. "The Archaeology of Wooden Boats." *Journal of the Institute of Wood Science* 7:1 (1975): 16–19. [ECU-214]

McKerrel, H., E. Roger, and A. Varsanyi. "The Acetone-Rosin Method for the Conservation of Waterlogged Wood." *Studies in Conservation* 17 (1972): 111–125. [ECU-407]

Mecklenburg, Marion, Charles Tumosa, and David Erhardt. "Structural Response of Painted Wood Surfaces to Change in Ambient RH." [ECU-773]

Merill, William. "Wood Deterioration Causes, Detection, and Prevention." *History News* (August 1974): 188. [ECU-329]

Merz, R.W. and G.A. Cooper. "Effect of Polyethylene Glycol on the Stabilization of Black Oak Blocks." *Forest Products Journal* 9:3 (1968): 55–59. [ECU-210]

Mikailov, A. "Conservation of a Thracian One-Log Boat." *ICOM Proceedings* (5th Triennial Meeting, Zagreb) (1978).

Mikolajchuk, E.A. et al. "Examination of Waterlogged Archaeological Oak Wood." *ICOM Proceedings* (Fremantle) (1987): 95–107. [ECU-200]

Moncrieff, A. "Review of Recent Literature on Wood, January 1960 to April 1968." *Studies in Conservation* 13 (1968): 186–212. [ECU-338]

Moss, S.T. "Preservation of Wood in the Sea." *The Biology of Marine Fungi*. Carrbridge: Carrbridge University Press, 1986: 355–365.

Muhlethaler, B., L. Barkman, and M. Noack. *Conservation of Waterlogged Wood and Wet Leather* (Paris: Editions Eyrothes) (1982). [ECU-369]

Munnikendam, R.A. "Conservation of Waterlogged Wood with Glycol Methacrylate." *Studies in Conservation* 18 (1973): 97–99. [ECU-339]

—. "Conservation of Waterlogged Wood Using Radiation Polymerization." *Studies in Conservation* [n.d].

Murdock, Lorne D. "Construction Details of a Tank for Polyethylene Glycol Treatment," *Proceedings*, Pacific Northwest Conference 2 (1976): 67–68. [ECU-228]

—. "A Stainless Steel Polyethylene Glycol Treatment Tank for the Conservation of Waterlogged Wood." *Studies in Conservation* 23 (1978): 69–75. [ECU-186]

Murphy, R.J. and D.J. Dickinson. "Wood Preservation Research—What have we learnt and where are we going?" *Journal of the Institute of Wood Science*. Vol. 14 No. 3, 1997. [ECU-647]

Nasser, A.F. "Some Thoughts on the Conservation of the Mombasa Hull." *Conservation of Waterlogged Wood* (1979): 45–49. [ECU-319]

Nemec, I. "Research in Waterlogged Wood—A Study in Conservation Methods." (1984).

Newlin, J.A. "The Religion of the Shrinkage and Strength Properties of Wood to its Specific Gravity." (USDA Technical Bulletin No. 676, Washington D.C.) (1919).

Nicholas, D.D. "Chemical Methods of Improving the Permeability of Wood." *Proceedings* (American Chemical Society Symposium, Series 43, I.S. Goldstein, ed., Washington D.C.) (1977): 33–46.

Nielsen, Hans-Otto. "The Treatment of Waterlogged Wood from the Excavation of the Haithbu Viking Ship." *ICOM Proceedings* (Grenoble) (1984): 299–312. [ECU-6]

Noack, M. "Some Remarks on the Processes Used for the Conservation of the Bremen Cog." *Proceedings* (ICOMOS Symposium on the Weathering of Wood, Ludwigsburg, Germany) [n.d.]: 89–97.

North, Neil A. "Determination of the Water Content of Acetone Solutions." *Studies in Conservation* 22 (1977): 197–198. [ECU-183]

O'Connor, S.A. "The Conservation of the Giggleswick Tarn Boat." *The Conservator* (1979): 36–38. [ECU-349]

O'Shea, C. "The Use of Dewatering Fluids in the Conservation of Waterlogged Wood and Leather." *Museums Journal* 71:2 (1971): 71–72.

Oddy, W.A. "Comparison of Different Methods of Treating Waterlogged Wood as Revealed by Stereoscan Examination and Thoughts on the Future of the Conservation of Waterlogged Boats." *Maritime Monographs and Reports* (National Maritime Museum, Greenwich, U.K.) 16 (1975): 45–49. [ECU-253]

Oddy, W.A., ed. *Problems of the Conservation of Waterlogged Wood.* (Maritime Monographs and Reports No. 16. Greenwich, U.K.: National Maritime Museum) (1975). [ECU-261]

—. "General Discussion, Day 2." *Maritime Monographs and Reports* (National Maritime Museum, Greenwich, U.K.) 16 (1975): 121–127. [ECU-260]

Organ, R.M. "Carbowax and Other Materials in the Treatment of Waterlogged Paleolithic Wood." *Studies in Conservation* 4 (1959): 96–105. [ECU-408]

Palfreyman, John W., George M. Smith, and Alan Bruce. "Timber Preservation: Current Status and Future Trends." *Journal of the Institute of Wood Science*. Vol. 14 No. 1, 1996. [ECU-648]

Pang, J.T.T. "The Treatment of Waterlogged Oak Timbers from a 17th Century Dutch East Indiaman, *Batavia*, Using Polyethylene Glycol." *ICOM-WWWG Proceedings* (Ottawa) (1981): 1–13.

—. "The Design of a Freeze-Drying System Used in the Conservation of Waterlogged Materials." *International Journal of Nautical Archaeology* 11:2 (1982): 105–111. [ECU-159]

Panshin, A.J. and de Zeeuw, C., Textbook of Wood Technology. Vol. 1, 3rd ed., (McGraw-Hill, N.Y., 1970).

Parrent, James M. "The Conservation of Waterlogged Wood Using Sucrose." *Proceedings* (14th Conference on Underwater Archaeology) (1983): 114–118. [ECU-70].

—. "The Conservation of Waterlogged Wood Using Sucrose." *Proceedings* (16th Conference on Underwater Archaeology) (1985): 83–95. [ECU-69]

—. "The Conservation of Waterlogged Wood Using Sucrose." *Studies in Conservation* 30:2 (1985): 63–72. [ECU-0308]

—. "The Conservation of Waterlogged Wood Using Sucrose. (Unpublished Manuscript) [n.d.]. [ECU-5]

Paterakis, A. "Conservation of Waterlogged Wood." *Archailogia* 8 (1984): 73–74.

Pearson, Colin. "The Use of Polyethylene Glycol for the Treatment of Waterlogged Wood—Its Past and Future." *Conservation of Waterlogged Wood* (1979): 51–56. [ECU-318]

Pearson, Colin, ed. *Conservation of Marine Archaeological Objects*. London: Butterworths Series in Conservation and Museology, 1987. [ECU-6]

Pendergast, M.F., J.V. Wright, and G.F. MacDonald. "Survey of Roebuck Prehistoric Village Site, Greenville County, Ontario." *ICOM-WWWG Proceedings* (1981): 189–200. [ECU-74]

Peterson, C.E. "New Directions in the Conservation of Archaeological Wood." *Archaeological Wood* (1990): 433–449. [ECU-293]

"Preserving and Conserving Wood." *Wooden Boat* Vol. 116, 1994, pp. 101–102. [ECU-774]

Plenderleith, H.J., and A.E.A. Werner. *The Conservation of Antiquities and Works of Art.* Second edition. London: Oxford University Press, 1971. [ECU-8]

Purdy, Barbara. "Survey, Recovery, and Treatment of Wooden Artifacts in Florida." *ICOM-WWWG Proceedings* (Ottawa) (1981): 159–169. [ECU-76]

Ramiere, Regis and Michel Collardelle, eds. *Waterlogged Wood: Study and Conservation.* (Proceedings of the 2nd ICOM-WWWG Conference, Grenoble) (1984). [ECU-9]

Ravindra, R., J.E. Davison, and R.H. La Fontaine. "The Storage of Untreated Waterlogged Wood." *Journal of the IIC-CG* 5:1/2 (1980): 25–31.

Reed, Sally A. "A Fortifying Preservative for Wood and Wood Fibers." *Curator* 9:1 (1966): 41–50. [ECU-77]

Resch, H. and B.A. Ecklund. "Permeability of Wood." *Forest Products Journal* 5 (1964): 199–206. [ECU-398]

Riess, Warren and Geoff Daniel. "Evaluation of Preservation Efforts for the Revolutionary War Privateer *Defence*." *International Journal of Nautical Archaeology*. Vol. 26 No. 4, 1997, pp. 330–338. [ECU- 649]

Rodgers, Bradley A. *The East Carolina University Conservator's Cookbook: A Methodological Approach to the Conservation of Water Soaked Artifacts.* Herbert R. Paschal Memorial Fund Publication, East Carolina University, Program in Maritime History and Underwater Research, 1992. [ECU-402]

Rosen, Howard. "High Pressure Penetration of Hardwoods." *Wood Science* 8:1 (1978): 355–363. [ECU-213]

Rosenquist, A.M. "The Stabilizing of Wood Found in the Viking Ship of Oseberg, Part II." *Studies in Conservation* 4 (1959): 62–72. [ECU-341]

—. "Experiments on the Conservation of Waterlogged Wood and Leather by Freeze-Drying." *MMR* (National Maritime Museum, Greenwich, U.K.) 16 (1975): 9–23.

—. "The Oseberg Find, Its Conservation and Present State." *Proceedings* (ICOMOS Symposium on the Weathering of Wood, Ludwigsburg, Germany) [n.d.]: 77–87.

Rowell, Roger M. and R. James Barbour, eds. *Archaeological Wood: Properties, Chemistry, and Preservation* (Advances in Chemistry Series 225, American Chemical Society, Washington D.C.) (1990). [ECU-292]

Saeterhaug, Roar. "Investigations Concerning the Freeze-Drying of Waterlogged Wood Conducted at the University of Trondheim." *ICOM-WWWG Proceedings* (Grenoble) (1984): 195–206. [ECU-81]

Sandstrom, Magnus, et al. "Deterioration of the Seventeenth-Century Warship *Vasa* by Internal Formation of Sulphuric Acid." Macmillan Magazines, Ltd, 2002. [ECU-601]

Sawada, M. "Some Problems of Setting PEG 4000 Impregnated Wood (Contraction of Impregnated PEG Solutions Upon Setting and Its Effect on Wood)." *ICOM-WWWG Proceedings* (Grenoble) (1984): 117–124.

—. "A Modified Technique for Treatment of Waterlogged Wood Employing the Freeze-Drying Method." *ICOM Proceedings* (Ottawa) (1981): 1–6.

—. "Conservation of Waterlogged Wooden Materials from the Nara Palace Site." *Proceedings* (International Symposium on the Conservation and Restoration of Cultural Property, Tokyo) (1977): 49–58.

Schaeffer, E. "Water Soluable Plastics in the Preservation of Artifacts Made of Cellulosic Materials." *Preprints* (5th Triennial ICOM Committee for Conservation Meeting, Zagreb) (1978): 1–16.

Scheizer, F., C. Houriet, and M. Mas. "Controlled Air Drying of Large Roman Timber from Geneva." *ICOM-WWWG Proceedings* (Grenoble) (1984): 327–338. [ECU-83]

Schmidt, J. David. "Freeze-Drying of Historic-Cultural Properties." *Technology and Conservation* 9:1 (1985): 20–26.

Schneiwind, Arno P. and Peter Y. Eastman. "Consolidant Distribution in Deteriorated Wood Treated with Soluable Resins." *Journal of the American Institute for Conservation.* Vol. 33 No. 3, 1994. [ECU-650]

Schweingruber, Fritz H. "Conservation of Waterlogged Wood in Switzerland and Savoy." *ICOM-WWWG Proceedings* (Ottawa) (1981): 99–106. [ECU-82]

Seborg, R.M. "Treating Wood with Polyethylene Glycol." *Diving into the Past* St. Paul: Minnesota Historical Society, 1964.

Seborg, R.M. and R.B. Inverarity. "Conservation of 200 Year Old Waterlogged Boats with PEG." *Studies in Conservation* 7 (1962): 111–120. [ECU-342]

—. "Preservation of Old, Waterlogged Wood by Treatment with Polyethylene Glycol." *Science* 136 (1962): 649–650.

Semczak, C. "Waterlogged Wood Preservation with Tletra-Ethyl-Orthosilicate." *Maritime Monographs and Reports* (National Maritime Museum, Greenwich, U.K.) 16 (1975): 151–152.

Sheetz, Ron and Charles Fisher. "Exterior Woodwork No. 4." *Preservation Tech Notes* (National Park Service) (1993). [ECU-518]

—. "Protecting the Woodwork Against Decay Using Borate Preservatives." *Preservation Tech Notes.* U.S. Department of the Interior, National Parks Service: Cultural Resources. [ECU-530]

"Shipworms and Gribbles: the Wooden Boat Eaters." *Wooden Boat* Vol. 111, 1993, pp. 66–68. [ECU-775]

"Shipworms: the Wood Boring Toredo Moves North." *Wooden Boat* Vol. 160, 2001, pp. 45–48. [ECU-776]

Siau, J.F. *Flow in Wood.* Syracuse: Syracuse University Press, 1971.

Siefert, Betty and Richard Jagels. "Conservation of the Ronson's Ship's Bow." *ICOM-WWWG Proceedings* (Grenoble) (1984): 269–292. [ECU-84]

Singley, Katherine R. "Design of a Large-Scale PEG Treatment Facility for the Brown's Ferry Vessel." *ICOM-WWWG Proceedings* (Ottawa) (1981): 1–13.

Skarr, Christen. *Water in Wood.* Syracuse, New York: Syracuse University Press, 1972.

Smith, C. Wayne. "Preservation of Waterlogged Wooden Buttons and Threads: A Case Study." *Historical Archaeology.* Vol. 36 No. 4, 2002. [ECU-651]

—. *Archaeological Conservation Using Polymers; Practical Applications for Organic Artifact Stabilization.* College Station: Texas A & M Press, 2003.

—. "Preservation of Waterlogged Wood Buttons and Threads: a Case Study." [ECU-777]

—. "The Re-Treatment of Two PEG-Treated Sabots." *Proceedings of the 7th ICOM-CC Working Group Wet Organic Archaeological Materials Conference.* Grenoble, France, 1998. [ECU-778]

—. "Treatment of Waterlogged Wood Using Hydrolyzable Multi-Functional Alkoxysilane Polymers." [ECU-779]

Smith, Herman A. "Conserving Iron Objects from Shipwrecks: A New Approach." *Proceedings* (Materials Research Society Symposia, No. 185, Vandiver, P.B., J. Druzik, and G.S. Wheller, eds.) (1991): 761–764.

Smith, Michael. "Wooden Artifacts." *Papers Presented at the 30th AIC Annual Meeting*, Miami Florida, 2002. [ECU-780]

Smith, W.B., W.A. Cote, R.C. Vasishth, and J.F. Siau. "Interactions Between Water-Borne Polymer Systems and the Wood Cell Wall." *Journal of Coating Technology* 57:729 (1985): 27–35. [ECU-0149]

—. "Study of Organic Interactions Between Wood and Water-Soluable Organic Solvents." *Journal of Coating Technology* 57:727 (1985): 83–90. [ECU-0148]

"Soft Hardwoods and Hard Softwoods." *Wooden Boat* Vol. 148, 1999, pp. 111–112. [ECU-781]

"Some Thoughts on Wood Chemistry." *Wooden Boat* Vol. 106, 1992, pp. 95–96. [ECU-782]

South, Stanley A. "Notes on Treatment Methods for the Preservation of Iron and Wooden Objects." (Unpublished manuscript, North Carolina Department of Archives and History) [n.d.]. [ECU-420]

Spriggs, James. "The Conservation of Timber Structures at York—A Progress Report." *ICOM-WWWG Proceedings* (Ottawa) (1981): 143–152. [ECU-0089]

Squirrell, J.P. and R.W. Clarke. "An Investigation into the Condition and Conservation of the Hull of the *Mary Rose*. Part I: Assessment of the Hull Timbers." *Studies in Conservation* 32:4 (1987): 153–162. [ECU-0220]

Stamm, A.J. "Shrinkage and Swelling of Wood." *Industrial and Engineering Chemistry* (1935): 401–406.

—. "Treatment with Sucrose and Invert Sugar." *Industrial and Engineering Chemistry* 29:7 (1936): 833–835. [ECU-218]

—. "Dimensional Stabilization of Wood with Carbowaxes." *Forest Products Journal* 6:5 (1956): 201–204. [ECU-414]

—. "Effect of Polyethylene Glycol on the Dimensional Stability of Wood." *Forest Products Journal* 9:10 (1959): 375–381. [ECU-211]

—. "Factors Affecting the Building and Dimensional Stablilization of Wood with PEG's." *Forest Products Journal* 14 (1964): 403–408.

—. *Wood and Cellulose Science*. New York: Ronald Press, 1964.

—. "Penetration of Cell Walls of Water Saturated Wood and of Cellophane by PEG's." *TAPPI* 51:1 (1968): 62.

—. "Penetration of Hardwoods by Liquids." *Wood Science Technology* 7 (1973): 285–296.

—. "Dimensional Changes of Wood and their Control." *Wood Technology* (American Chemical Society Symposium, Washington D.C.) (1977): 115–140.

Stamm, A.J. and R.M. Seborg. "Minimizing Wood Shrinkage and Swelling by Treating with Synthetic Resin-Forming Materials." *Industrial and Engineering Chemistry* 28 (1936): 1164–1169.

Stark, Barbara. "Waterlogged Wood Preservation with Polyethylene Glycol." *Studies in Conservation* 21:3 (1976): 154–158. [ECU-335]

Sturla, A. "Data Processing as a Decision-Making Tool in the Treatment of Waterlogged Wood." *ICOM-WWWG Proceedings* (Grenoble) (1984): 161–165.

Suthers, T. "The Treatment of Waterlogged Oak Timbers from Ferriby Boat III Using Polyethylene Glycol." *Maritime Monographs and Reports* (National Maritime Museum, Greenwich, U.K.) 16 (1975): 115–119. [ECU-247]

de Tassigny, C. "The Suitability of Gama Radiation Polymerization for Conservation Treatment of Large Size Waterlogged Wood." *Conservation of Waterlogged Wood* (1979): 77–83. [ECU-314]

Tesoro, F.O. and E.T. Choong. "Relationship of Longitudinal Permeability to Treatability of Wood." *Holzforschung* 30:3 (1976): 91–96.

Tesoro, F.O., E.T. Choong, and O.K. Kimbler. "Relative Permeability and the Gross Pore Structure of Wood." *Wood and Fiber* 6:3 (1974): 226–236.

Tilbrooke, D.T. "The Deterioration of Wood in a Benthic Environment." *Proceedings* (First Southern Hemisphere Conference on Maritime History) (1978): 147–150.

Titus, Larry. "Conservation of Wooden Artifacts." *ICOM-WWWG Proceedings* (Ottawa) (1981): 153–158. [ECU-93]

Tomashevich, G.N. "The Conservation of Waterlogged Wood." *Problems of Conservation in Museums* (London: George Allen and Unwin, Ltd.) (1969): 165–186.

Tuck, James A. "Conservation of Waterlogged Wood at a 16th Century Whaling Station." *ICOM-WWWG Proceedings* (Ottawa) (1981): 171–175. [ECU-94]

"Two PEG's Are Better than One." *Chemistry in Britain* (1987). [ECU-107]

Kleingardt, Birgitta, ed. "Preserving the *Vasa*." *Wasa*. Stockholm: Royal Printing Office PA Norstedt and Soner, 1969. [ECU-579]

U.S. Borax. "Tim-Bor." *U.S. Borax Service Bulletin 200 & Speciman Label*. [ECU-529]

Van Der Heide, Gerrit. "Considerations Regarding the *Amsterdam* as an Historical Monument." *Conservation of Waterlogged Wood* (1979): 99–105. [ECU-315]

—. "A Piece of History in Conservation of Waterlogged Wood." *Conservation of Waterlogged Wood* (1979): 17–24. [ECU-0302]

—. "The Present Situation with Respect to Shipwrecks." *Conservation of Waterlogged Wood* (1979): 9–16. [ECU-303]

Van Geersdale, P.C. "Note on the Direct Application of Plaster of Paris to Waterlogged Wood." *Studies in Conservation* 20:1 (1975): 35.

—. "Plaster Moulding of Waterlogged Wood." *Maritime Monographs and Reports* (National Maritime Museum, Greenwich, U.K., National Maritime Museum, Greenwich, U.K.) 16 (1975): 109–111. [ECU-249]

Vikhrov, V.E. et al. "Pickling Old Boats in Alcohol." *New Scientist* 63:904 (1974): 27. [ECU-111]

Van Dienst, E. "Some Remarks on the Conservation of Wet Archaeological Wood." *Studies in Conservation* 30 (1985): 86–92.

de Vries-Zuiderbaan, Louis H., ed. *Conservation of Waterlogged Wood* (International Symposium on the Conservation of Large Objects of Waterlogged Wood. The Hague: Netherlands National Cornmission for UNESCO) (1979). [ECU-304]

Wachsmann, Shelley. "The Galilee Boat—2000 Year Old Hull Recovered Intact." *Biblical Archaeological Review* 14:5 (1988). [ECU-346]

Wallace, Patrick F. "The Survival of Wood in Tenth to 13th Century Dublin." *ICOM-WWWG Proceedings* (1984): 81–87. [ECU-95]

Wang, Y. and A.P. Schiewind. "Consolidation of Deteriorated Wood with Soluable Resins." *Journal of the American Institute for Conservation* 34:2 (1985): 77–91. [ECU-96]

Ward, Clare, Derrick Giles, Dean Sully and David John Lee. "The Conservation of a Group of Waterlogged Neolithic Bark Bowls." *Studies in Conservation.* Vol. 4 No. 4, 1996, pp. 241–249. [ECU-652]

Wardrop, A.B. and G.W. Davies. "Morphological Factors Relating to the Penetration of Liquids into Wood." *Holzforschung* 15:5 (1961): 129–141.

—. "Waterlogged Wood: The Recording, Sampling, Conservation and Curation of Structural Wood." *Proceedings* (Conference sponsored by WARP and English Heritage) (1990).

Watson, Jacqui. "Research into Aspects of Freeze-Drying Hardwoods Between 1982–1984." *ICOM-WWWG Proceedings* (Grenoble) (1984): 213–218. [ECU-97]

—. "The Application of Freeze-Drying on British Hardwoods from Archaeological Excavations." *ICOM-WWWG Proceedings* (Ottawa) (1981): 237–242. [ECU-98]

Williams, D. "Identification of Waterlogged Wood by the Archaeologist." *Science and Archaeology* 14 (1975): 3–4.

de Witte, E., A. Terve, and J. Vynckier. "The Consolidation of the Waterlogged Wood from the Gallo-Roman Boats of Pommeroeul." *Studies in Conservation* 29 (1984): 77–83. [ECU-124]

Young, Gregory S. "PEG Localization within the Structure of Waterlogged Wood." *Proceedings* (IIC Congress, Poster Session, Washington D.C.) (1982).

Young, Gregory S. and I.N.M. Wainwright. "A Study of Waterlogged Wood Conservation Treatments at the Cellular Level of Organization." *ICOM-WWWG Proceedings* (Ottawa) (1981): 107–116. [ECU-101]

Young, Gregory S. and Ritchie Sims. "Microscopical Determination of Polyethylene Glycol in Treated Wood—The Effect of Distribution on Dimensional Stabilization." *ICOM Proceedings* (Fremantle) (1987): 109–140. [ECU-201]

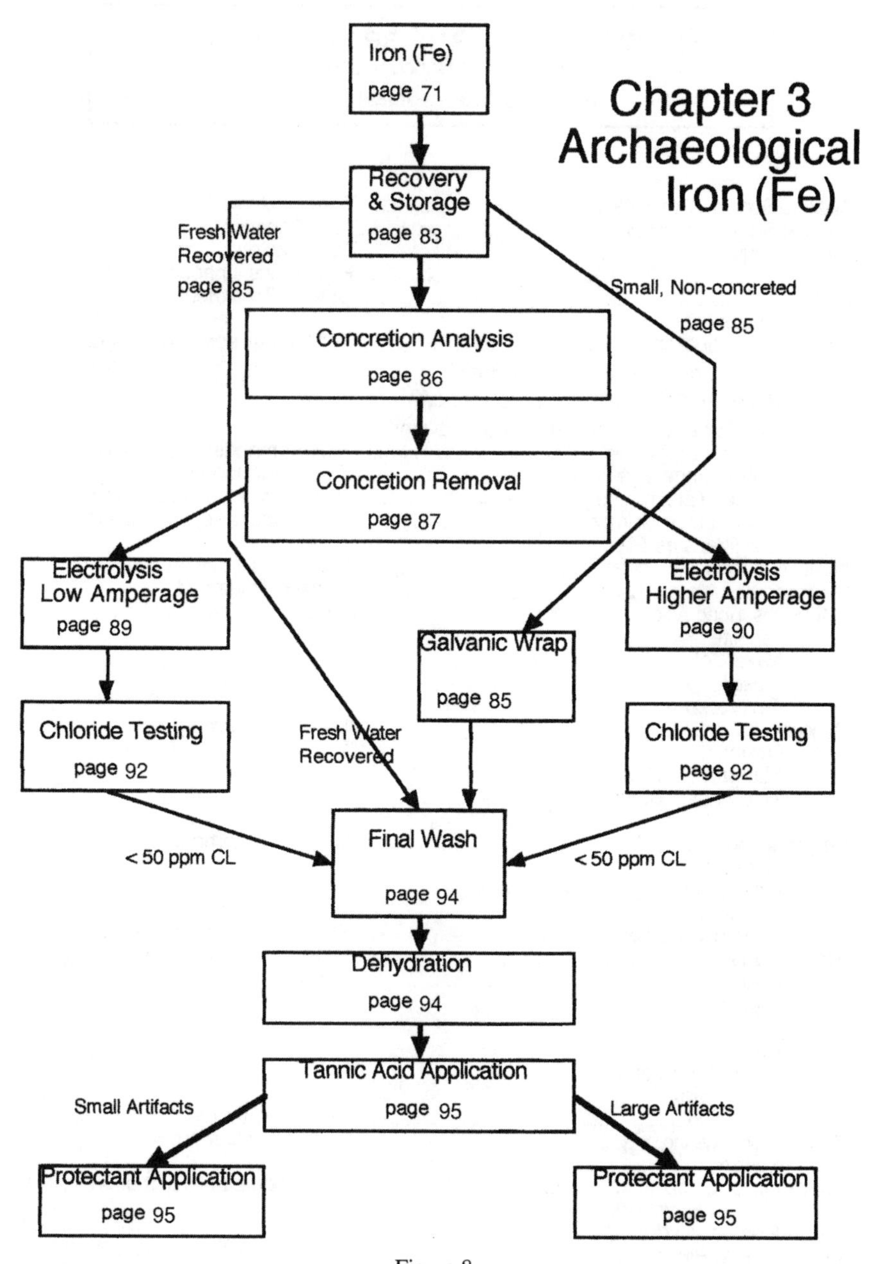
Iron (Fe)
page 71
Chapter 3
Archaeological
Iron (Fe)
Recovery
& Storage
page 83
Fresh Water
Recovered
page 85
Small, Non-concreted
page 85
Concretion Analysis
page 86
Concretion Removal
page 87
Electrolysis
Low Amperage
page 89
Electrolysis
Higher Amperage
page 90
Galvanic Wrap
page 85
Chloride Testing
page 92
Fresh Water
Recovered
Chloride Testing
page 92
< 50 ppm CL
Final Wash
page 94
< 50 ppm CL
Dehydration
page 94
Tannic Acid Application
page 95
Small Artifacts
Large Artifacts
Protectant Application
page 95
Protectant Application
page 95

Figure 8.

Treatments for Archaeological Iron
Represented in Conservation Literature

Primary source (Clarke and Blackshaw, 1982; 18 - 20)

Alkaline Sodium Dithionite...Toxic Factors
Alkaline Sulfite Reduction...Toxic Factors
Amonia (Liquid)...Toxic Factors
Annealing...High temps. changed micro-structure
EDTA..Surface Treatment
    Detarex C
Electrolytic Reduction...Works well, some toxic electrolytes
    High Current Density .1 amp/cm^2
    Moderate Current Density .05 amp/cm^2
    Low Current density.001 - .005 amp/cm^2
    Low amperage/ light electrolyte......................Recommended (page 89)
    Higher amperage/ light electrolyte..................Recommended (page 90)
Galvanic Wrap (Electrochemical Cleaning)................,Recommended (page 85)
Hydrogen Reduction Furnace.......................................Fast, expensive, fire hazard
    Low Pressure Plasma
In Situ Galvanic Reduction...Stabilization, not treatment
Mechanical Surface Cleaning.......................................Surface Treatment
Orthophosphoric Acid...Surface Treatment
    Jenolite
    Naval Jelly
    Trustan
    Ferroclean
Phosphoric Acid..Surface Treatment
    Ospho
    Manganese Phospholene
    Zinc Phospholene

Sequestering Agents..Surface Treatment
    Nervanaid
    Metaquen
    Dequest
Sodium Carbonate Wash..Slow, non-chlorinated artifacts
Soxhlet Washing...Works on small friable artifacts
Stripping Acids...Surface Treatment
    Oxalic
    thioglycollic
    Citric
        Modalene
        Biox
Surface Cleaning..Surface Treatment
Tannic Acid Coating..Surface Treatment
    Fertan
Washing...Slow, some toxic procedures
    Lithium Hydroxide
    Sodium Hydroxide
Water Stream Rinsing...Slow, need running stream

Figure 9.

Chapter 3

Archaeological Iron (Fe)

TYPES OF IRON—THEORY

Iron is the most plentiful and functional metal used by humans. Though iron from meteorites caught the eye of early Bronze Age people, they could not smelt it from ore until much later. The Hittites in Asia Minor used quantities of iron as early as 1400 BC and iron use became wide spread by 1000 BC. Unlike many other metals, iron does not occur in pure form naturally. It is most often found as a mineralized oxide, the two most plentiful being hematite and magnetite. The greatest problem faced by the modern conservator is how to keep iron and its alloy steel from reverting to its more stable oxide forms through corrosion.

Iron is produced in three basic types; pig, cast, and wrought. These three types of iron reflect in descending order the carbon content, or in ascending order, the refinement of the metal. Steel is an alloy of carbon and iron and does not greatly impact the archaeological record until its mass production in the mid 19th century. But steel could be produced in small quantities for blade edges and springs early in the history of iron manufacturing.

Pig iron is produced from iron ore that is melted and reduced (technical term that means to gain electrons) in a furnace. The term pig iron is an illusory metaphor for the image of suckling piglets, since furnaces used to produce the molten iron typically ran the metal by gravity into molds surrounding its base. Furnaces became more efficient over time undergoing development from hearth and shaft furnaces of medieval times to blast furnaces in the 14th century when it was found that the introduction of large quantities of air forced into the furnace heated the ore faster and reduced it more efficiently. Blast furnaces of all types invariably combine iron ore, coke or charcoal, and limestone. Hot air is forced into the blast furnace igniting the coke or charcoal that in turn gives off carbon monoxide. The carbon monoxide picks up oxygen from the hot iron ore becoming carbon dioxide, reducing the ore to metal. The calcium from the limestone acts as a flux or catalyst to lower the melting temperature of the impurities in the iron ore to that of the iron. This allows the melted iron and impurities to be cast into pigs at the bottom of the blast furnace.

Pig iron has a carbon content of 3% to 6%, is very brittle and is not useful for tools. Pig iron is an unlikely, though not unheard of find on an archaeological site. It was sometimes transported in ships for refinement and also used as ballast,

Figure 10. An 18th century cast iron pipkin from a fresh water site. Note the spalling corrosion on the lip indicating the artifact was found half buried on its side, with the buried portion exposed to anaerobic sulfate reducers and methanogens. Photograph by Chris Valvano.

most often in warships. Pig iron is generally black in color with a rough surface. All iron can be polished to silvery base metal but it rarely if ever appears this way in an artifact.

Cast iron is a refined version of pig iron. It contains a high carbon content of 2% to 6% and remains unworkable and brittle. Yet it can be cast into useful objects such as skillets, pots, piping, stoves, ranges, cannon, machine parts, and cannon balls. Cast iron is very hard but remains brittle due to its micro-crystalline structure. It was not unheard of for cannon shot to shatter on impact with ships or fortifications and certainly most plumbers still realize today that the easiest way to remove cast iron piping is to shatter it with a hammer. Cast iron, like pig iron, is normally black in color and can have a rough or smooth texture depending on the mold in which it was formed. Cast iron corrodes in layers from the surface inward and may crack and exfoliate. The key to cast iron corrosion is that the metal cools onto a carbon graphite matrix after casting, giving it a crystalline structure. As the iron corrodes and migrates from the matrix the artifact will retain its shape but the outer layer may only be very soft graphite. Conserving the archaeological surface in this case may mean preserving the soft graphite outer layer.

Wrought iron is the most refined version of iron, processed from high quality pig iron. This iron is purified and poured over melted sand in order to produce bloomery iron or blooms, as the ingots are called. The silicate slag (sand) introduced into the wrought iron allows this material to be worked and formed into tools and fasteners in a forge. Wrought iron, like steel, does not show up in great quantities in the archaeological record until Henry Cort used both grooved rolling machines and the puddling process to refine large quantities of the metal beginning in 1784.

Wrought iron is highly malleable and does not shatter. It is nearly pure with less than .2% carbon content and contains silicate slag in the form of stringy intrusions that run with the grain of the worked metal. Wrought iron is normally

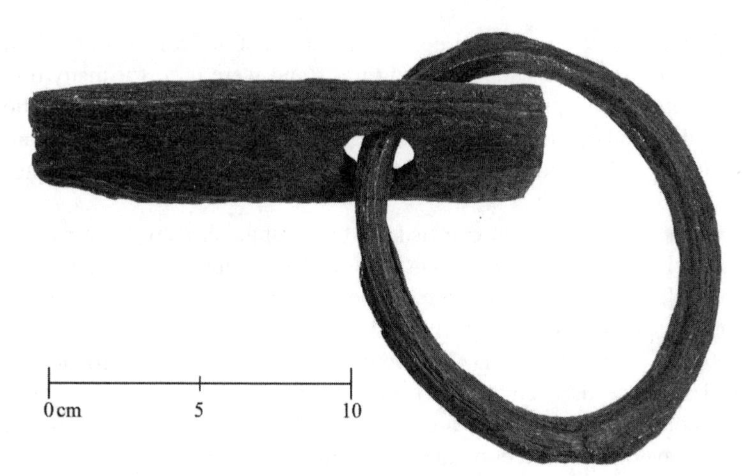

Figure 11. Wrought iron log staple of the 19th century. Note the grainy appearance of wrought iron as it corrodes along the silicate slag intrusions. Photograph by Chris Valvano.

brown or dark brown in color. Archaeological wrought iron will have a grainy wood-like appearance as it corrodes along the silicate slag intrusions. Wrought iron will be formed into any worked utilitarian objects that will see hard use, such as fasteners, nails, and anchors.

Steel could be produced in small amounts for edged weapons early in the history of iron working, and after the mid 18th century as crucible steel (in which, measured amounts of carbon were worked into wrought iron). Unlike iron, steel springs back when bent and is lighter and stronger than iron. But steel was not mass-produced before Englishman Henry Bessemer patented his steel making process in 1856. In this process molten iron from a blast furnace is poured into a helmeted bucket mechanism called a converter. Carbon, and manganese were added to the iron in the converter and air pumped through it to burn off impurities. The converter purified the iron and controlled the carbon content to produce steel. In the last quarter of the 19th century, German brothers William and Frederick Siemens and French counterparts Pierre and Emile Martin improved Bessemer's process, making it more efficient and productive. The Siemens-Martin open-hearth furnace remained the dominant producer of steel throughout the last part of the 19th century.

Steel can be silvery metallic in color after production. It corrodes much faster than iron and, left to its own devices, will soon take on an iron oxide patina of brown, red, or orange. Steel contains a .5% to 3% carbon content and can be alloyed to many other metals such as nickel, cobalt, manganese or chromium. These alloy steels have unique properties and are used for different purposes. In general steel corrodes in large flakes that exfoliate from the surface of the artifact. Archaeological steel includes springs, armor, swords, and the edges of utilitarian tools such as axes.

Though pig, cast, and wrought iron plus steel are primarily made of the element iron, they have widely varying traits and were used for many different purposes throughout history. Not surprisingly the deterioration of these alloys is also very different on the macro and microscopic level. Artifacts of these materials will, of course, also deteriorate and behave in widely variable manners and must be conserved accordingly.

Artifacts made of iron can usually be distinguished by their dark brown to black color, hardness, and weight. Iron and steel are also magnetic and will carry and electric current. Unlike more noble metals iron readily corrodes, with the corrosion products moving outward into the environment to precipitate on encountering a more basic pH. The precipitated iron corrosion product (mostly FeO(OH), orange rust) cements all nearby debris to the artifact. In aerated moist soil this concretion becomes a brown amorphous mass complete with attached sand and gravel. This concretion may have little resemblance to the artifact hidden inside. Underwater the concretion process is more complex but will result in the complete concealment of the iron object within a calcium carbonate shell (see section on concretion formation).

Archaeologists and conservators will, through experience, become familiar with the normal uses of these types of materials. It would be impossible for instance to make cast iron nails or horseshoes, for they would shatter in short order. On the other hand, complex molded items are impossible to make of wrought iron, as it cannot be poured into a mold. Mold marks, and decomposition pattern are the best indicators for iron type. Some other artifacts such as metal buckets and so called, "tin cans," are also made almost entirely of iron, and later steel.

CORROSION OF IRON—THEORY

Iron is most stable in its natural form as an oxide. If this oxide is reduced and becomes a nearly pure metal, it will corrode in order to return to its most stable state—an oxide. Metallurgists through time have understood this and by the end of the19th century were able to communicate this general corrosion theory in the formula:

$$4Fe^0 + 2H_2O + 3O_2 \rightarrow 4FeO(OH)$$

Simply stated this says that iron plus water plus oxygen will turn to rust (ferrous oxy-hydroxide). This was easily observable where-ever iron was used but failed to explain some other observable consequences of using iron, particularly iron fasteners in ships. For example, when ship builders first resorted to using copper sheathing on the bottoms of their wooden vessels (HMS Alarm 1758–1762) to protect them from toredo worms and shell fish, they noticed that the iron fasteners began to corrode at an alarming rate. This mysterious corrosion led to the near destruction of the experimental vessels. From that point onward all fasteners inside ships that were to be copper sheathed were also made of copper or copper alloy.

Later shipbuilders noticed that iron fasteners, even when painted and protected from oxygen, could mysteriously waste away inside their protection. And of course all iron users noticed that iron corroded faster in a salt water or seaside environment. These observations eventually led to our present understanding of galvanic coupling (two metals in contact), differential aeration corrosion (corroding metal does not necessarily need to be exposed to oxygen and water to corrode), and the deleterious effects of the chloride anion on metals.

All of these forces play a role in the deterioration of iron. All archaeologically recovered artifacts, from land or sea sites, must be protected from salt intrusion, galvanic coupling, and differential aeration. And to a certain extent conservators will have to reverse these processes in order to conserve and stabilize iron.

It is now understood that the simple general corrosion formula is actually two processes. Corrosion or oxidation is the movement of electrons both within and between metals and the freeing of metallic ions (charged particles of a metal) while reduction is the collection of electrons. All corroding metals create a battery with a positive pole and a negative pole. At the positive side (+) of the battery, or the Anode, the reaction can be described by the formulas:

Anode +

$$Fe^0 \rightarrow Fe^{+2} + 2e^-$$

$$O_2 + 4e^- + 2H_2O + 2Fe^{+2} \rightarrow 2FeO(OH) + 2H^+$$

Stated simply this means that iron under proper circumstances will give off electrons. In so doing iron atoms become charged ions and charged ions will migrate. The migrating iron ion will encounter oxygen, energy, and water and turn to rust while also producing hydrogen ions, or acid. In short, corroding iron produces rust, acid, and electricity.

In order for a direct current battery to work, the circuit must be completed. In other words the electrons given off at the positive pole must be used. This happens at the cathode.

Cathode –

$$O_2 + 2H_2O + 4e^- \rightarrow 4OH^-$$

The electrons given off by the corroding iron reduce oxygen and water to produce hydroxyl ions, or a base. Paint and other electrically passive substances restrict the movement of electrons, polarizing the electrical transfer between the anode and the cathode, slowing the reaction.

This new understanding of corroding metals helps explain one of the 19th century corrosion mysteries, that of differential aeration. It can be seen from the formulas that the anode and cathode of a reaction do not have to occur in the same place, they only have to be connected by an electrically conductive path for electrons. The corroding metal in a steel ship, for instance, can be separated from the cathode or that place where the oxygen and water are reduced, by a great distance. This also implies that so long as any portion of an iron artifact is exposed to oxygen and water, it will corrode—and not necessarily at the area of exposure. A diver chipping paint from the surface of a shipwreck may cause corrosion of metals deep inside the wreck, or an iron nail may corrode inside an object because its head is exposed to oxygen and water.

GALVANIC COUPLING—THEORETICAL

In essence, therefore, corrosion is an electron transfer that will continue so long as electrons are produced at the anode and used at the cathode. When dissimilar metals come into electrical contact they form what is called a galvanic couple. In this couple, one metal will become an anode and corrode at an accelerated pace, while the other becomes a cathode and is protected. Galvanic coupling is another of the 19th century corrosion mysteries explained by a transfer of electrons.

All metals have an average corrosion potential (E Corr) that can be measured in milli-volts of electricity, but each type of metal has a different E Corr. The measured E corrs correspond to the electrons given off when a small amount of metal ionizes. Some metals corrode more readily than others. Gold for instance does not corrode as easily as iron and is considered a more noble metal. If two dissimilar metals come into electrical contact or perhaps are manufactured in contact, such as a brass shoe buckle with an iron pin, or a brass bell with an iron clapper (see photograph), the less noble metal will begin to donate electrons to the more noble metal. By definition the less noble electron donor (in the example iron) will corrode and oxidize while the more noble metal (brass—alloy of copper) will be reduced or gain electrons. Since reduction is used to purify a metal from ore it will also discourage oxidation in the noble metal. In other words, the more noble

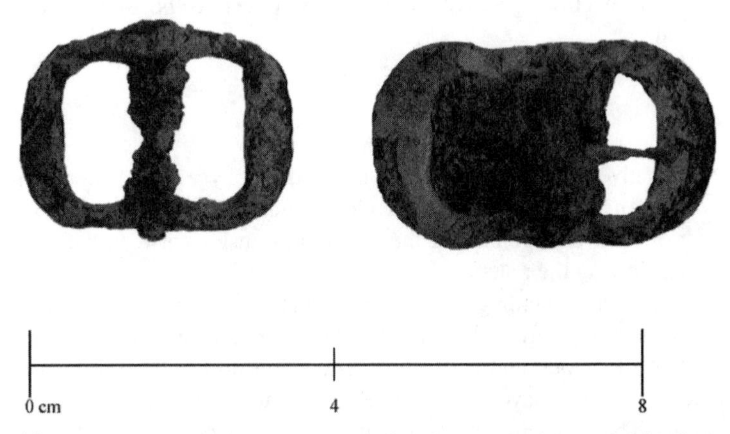

Figure 12. Brass buckles with iron pins. Note the iron has completely corroded forming a central concretion while the brass outer buckle is in relatively good condition because of the galvanic coupling. Photograph by Chris Valvano.

metal will be protected and may look to be in pristine condition though it has been buried, or submerged, for hundreds of years.

CONCRETION FORMATION—THEORY

Non-concreted iron is rarely recovered from an archaeological site on land or in the sea. Fresh water sites in lakes and rivers seem to spawn the fewest concretions but they can still form at times when calcium carbonate precipitates on an iron object. Even hydro-thermal rivers have been found to contain artifacts encompassed in a concreted material know as armoring. Though these concretions begin as an electro-chemical process initiated by iron corrosion they invariably become a biological system as well. Armoring and sea bed concretions are both colonized by sulfate reducing bacteria and methanogens.

Concretions recovered from the earth invariably contain sand and other hard rock cemented together in a ferrous oxide and ferrous carbonate (cementite) matrix. J.M. Cronyn refers to these concretions as a red/brown mass that are often, "no longer recognizable as a particular object (Cronyn, 1990; 179)." These dry recovered concretions are most often found surrounding wrought iron, but can also form nodules on cast iron.

The actual concretion process on iron in the earth is not well represented in literature but likely mimics in many ways the formation process of concreted artifacts abandoned in anaerobic seabed environments. However, there are some important differences.

Concretions begin to form on earth interred iron artifacts almost immediately when there is available oxygen and water—both plentiful in most soils. The artifact initiates this process by giving off electrons, iron ions, and acid. Of utmost concern here is the microscopic artifact/soil interface. Here, carbonates in the soil that surrounds the artifact dissolve from the acid produced by the corroding metal.

Figure 13. Typical brown mass concretion of terrestrially recovered iron nail. The concretion consists of carbonates, sand, and iron corrosion products. Most of the iron in the nail has already migrated into the concretion. Photograph by Chris Valvano.

The carbonates and ferrous ions are then free to migrate away from the immediate vicinity of the object and soon encounter a more basic pH in the surrounding soil. The higher more alkaline pH allows the iron ions and carbonates to precipitate out as ferrous oxy-hydroxides (orange colored rust) and ferrous carbonates. In this way the precipitates begin to build a shell around the artifact, a shell that cements together all of the sand particles and small rocks and pebbles nearby. This budding concretion becomes the cathode on which the oxygen and water is reduced by the electrical charge escaping from the artifact. The artifact produces a charge that is transported by migrating ions, ferrous and hydrogen cations, and various earth salt anions including chlorides. These anions will penetrate into the body of the artifact and need to be removed from the object if it is to be stabilized during conservation.

The interior of the concretion with its acid pH makes for an ideal corrosive environment. Unlike concretions formed in the sea this amorphous red/brown mass does not seem to be electrically passive and is porous to both oxygen and water, meaning the corrosion process is free to continue. Since the electrons produced by the artifact are used as fast as they are produced the system continues to corrode the artifact. Migrating acid, iron ions, and carbonates expand the concretion outward at the expense of the artifact.

However, another process seems to be at work on the earth interred concretion, a destructive dry cycle. The dry cycle must occur whenever the soil dries and there is little reducible water available. In the dry cycle the acid and corrosion products produced by the artifact cannot migrate so they build pressure under the concretion. Cracks in the concretion radiate from the artifact and will continue until it begins to spall and break apart. The dry cycle will also flake off large pieces of the artifact inside the concretion as corrosion products expand in cracks and fissures. Evidence shows that many earth-interred concretions show repeated cracks that later fill with corrosion product. It should be understood that the mineralized versions of iron occupy more space physically than metallic iron, therefore, corrosion products will always expand. The dry cycle is extremely destructive to artifacts in the soil and for that matter, for artifacts brought into the laboratory without concretion removal and deep rinsing of anions.

Some iron artifacts buried deeply in anaerobic soils are subject to attack by sulfate reducing and methanogenic bacteria. Concretions are not formed around these artifacts, instead they are distinguished by odorous black iron sulfides. These artifacts will be badly corroded even though there is no oxygen present. The sulfate reducers and methanogens metabolize electrons given off by the artifact so the artifact does not Polarize, but continues to give off electrons (see section on ocean concretions). Other soil conditions are not conducive to concretion formation but do seem to promote electrically passive protective layers over iron artifacts buried in wet organic soils rich in phosphate (Cronyn, 1990; 181). And organic iron tannic compounds may also help protect non-concreted iron in swampy conditions.

The concretions formed on terrestrially recovered artifacts can be extremely hard and difficult to remove particularly if they are composed of sand. The corrosion product and corbonate matrix adheres to the artifact and may flake off parts of the

Figure 14. Wrought iron log staple undergoing dry cycle decomposition. Photograph by Chris Valvano.

artifact as it is removed. The dry cycle system of concretion formation in the earth may also have contributed to the actual loss of the surface of an iron artifact. Concretions produced in the ocean do not normally cling so tightly to the artifact, are frequently easier to remove, and when they are removed the archaeological surface of the iron artifact is usually intact.

OCEANIC CONCRETION FORMATION—THEORY

The best studies conducted thus far on concretion formation in the ocean come from Neil North and Ian MacLeod of the Western Australia Maritime Museum (Pearson, 1987; 77–78). In their research North and MacLeod concluded that because iron artifacts lying in the ocean are not toxic to living organism they are colonized in short order, first with fouling assembledges, then by coralline algae, and finally within a few years by a layer of hard coral. Unlike earth formed concretions, ocean concretions are almost exclusively formed of calcium carbonate, are fairly non-porous, and inhibit ion exchange.

Nonetheless, ions continue to diffuse into and out of the concretion albiet at a slowed pace. Iron Fe^{+2}, Fe^{+3}, and H^{+} cations move to the concretion while the ever-present chloride Cl^{-} anions, from the salt in ocean water, move to the artifact. Some iron artifacts recovered from the ocean may contain up to 10% of their weight as chlorides. As the iron ions and acid diffuse outward and encounter more alkaline conditions they precipitate in the form of oxides, hydrated oxides, and hydroxy chlorides. Some iron ions begin to take the place of the calcium in

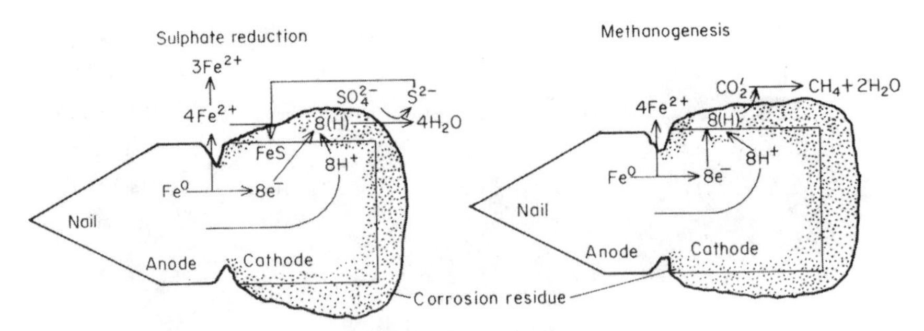

Figure 15. Sulfate reduction and Methanogenesis in an anaerobically interred iron nail (Rodgers, 1989: 337).

the calcium carbonate concretion forming cementite, a much harder version of the original concretion.

While the surface of an artifact invariably becomes a living ecosystem the same can be said of the interior of the concretion. At some time in its formation, anaerobic sulfate reducing and methanogenic bacteria colonize the interior of the concretion. These creatures thrive in acidic conditions and metabolize the electrons given off by the artifact and produce, as byproducts, iron sulfides and methane gas. Because the electrons from the corroding artifact are no longer hindered, iron ions are produced at an increasing rate.

Over time the concretion and corrosion process will continue until the artifact disappears leaving a hollow cast of itself in the concretion. Since there is no dry cycle in the ocean the concretion will remain relatively intact as a replica of the original artifact.

Anaerobic concretion formation in the ocean follows a slightly different course than that on the sea floor. Anaerobic concretions form when a buried iron object produces iron ions and acid. The micro-environment near the surface of the artifact becomes supersaturated with dissolved calcium carbonate due to the acid present. As the supersaturated water moves away from the artifact it comes into contact with more alkaline conditions and the calcium carbonate precipitates out, bonding to it all of the nearby sand, coral, rock, and artifacts. Concretions formed in the benthic, or ocean bottom environment, are easily distinguishable from those formed on the sea floor because of the attached pebbles and rock, and the fact that there is no coral or shellfish attached to the outer layer.

Inside, just as with the sea floor formed concretion, the sub bottom concretion will be colonized by sulfate reducing and methanogenic bacteria. Complete disintegration of this artifact is assured over time, but it too will leave a cementite mold beneath the sea bottom.

This discussion of concretions cannot be concluded without emphasizing the fact that an understanding of how concretions form and what processes take place in them is of vital importance to both a conservator and an archaeologist. The conservator needs this information to gauge the effects of time on an artifact. Iron stabilization can be a long process and concretion removal is only a small

Figure 16. Hollow concretion from an ocean site. The iron from the artifact has corroded and moved into the concretion leaving a hollow mold of what it once looked like. Photograph by Chris Valvano.

part of it. The amount of chlorides absorbed and the condition of the artifact will determine its overall treatment time. Conservators can often gauge the condition of an artifact by the amount of iron in the concretion. In a similar way archaeologists should be able to say a great deal about site formation process by simply looking at if, and how, the iron at a site is concreted.

POURBAIX DIAGRAM—THEORY

The final section of this theory discussion will be used to describe a tool that can be used by both conservators and archaeologists to characterize a site. The Pourbaix Diagram is a simple illustration that demonstrates at a glance the type of site conditions present overall. It can also be used at a more focused level to describe the conditions found at or near a single artifact or area on an artifact. The Pourbaix is set up in different ways, but it is always a plotting of pH vs. Eh. pH, or partial hydrogen, is a measurement of acidity, and denoted by numbers 1–14 indicating the negative logarithm of the hydrogen ion concentration. pH values run from 1 to 7 on the acidic side, and from 7 to 14 on the basic or alkaline side. A pH of 7 is neutral. Eh is the Redox Potential. It is the amount of electron pressure (voltage) being given off by artifacts on a site. Both Eh and pH can be measured by meter or simpler means such as titration strips in the case of pH. The theoretical limit for a metal artifact that is giving off electrons is about 1.5 volts, so the scale for Eh runs to plus or minus 1.5 volts with 0 as neutral.

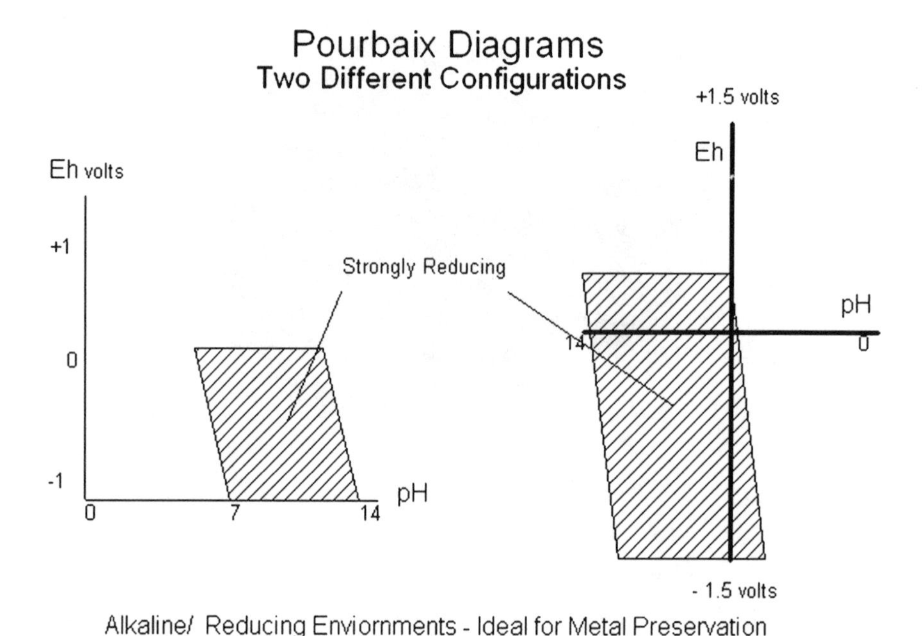

Figure 17. Two different configurations of Pourbaix Diagrams demonstrating areas within the Eh/pH framework that are conductive to artifact preservation.

Interpretation of the diagram can be accomplished by looking at the zones that are formed. The most beneficial zones for artifact preservation are alkaline/reduction zones, while most artifact deterioration will take place in the acid/oxidizing zones. The other zones alkaline/oxidizing and acid/reduction offer a mixed bag environmentally, preservation will depend on the artifact and its individual circumstance.

ELECTROLYSIS—THEORY

Various methods have been devised over the years to mitigate and even reverse the corrosion process in archaeologically retrieved iron artifacts. Many chemical, electrochemical, and heat treatments are successful in reducing the iron and stabilizing the artifacts. Yet, none of these methods are as cheap, easy to accomplish, or can offer ways of stabilizing iron without health or pollution risks as electrolysis or electrochemical cleaning in the form of the galvanic wrap.

The principle behind electrolysis or electrolytic reduction is simple. A direct current is applied to an artifact and a sacrificial anode in an electrolyte. The artifact is grounded to the negative terminal to become the cathode while the sacrificial anode is connected to the positive pole of the direct current source. This artificially reverses the electron flow in the artifact and reduces it. As the artifact is reduced the corrosion products on it are also reduced. Though electrolysis cannot supply the energy needed to turn the corrosion products back to elemental iron, they will reduce to magnetite, a much more stable form of mineralized iron.

Corroding metal oxides invariably expand outward to take up more space than the original metal. In contrast, when a metal is reduced and heated to refine it to a pure metal, it shrinks. While the corrosion products are electrically reduced (collect electrons), they are also physically reduced in size. That is, the corrosion products change from various oxides and hydroxides (RUST) into magnetite and in so doing become physically smaller. This reduction in size of the corrosion product opens up the surface of the artifact to the rinsing action of the electrolyte. The chlorides and other anions from the environment that have moved into the artifact begin to leach out of the pores, micro cracks, and interior of the artifact out into the electrolytic solution since their pathways are no longer blocked by corrosion product.

Electrolysis in iron, therefore, is nothing more than a facilitated rinse. The surface of the artifact will turn dark with stable magnetite while the harmful anions gradually wash from the object.

ARTIFACT STORAGE—METHODOLOGY

ALL recovered iron artifacts are actively corroding. While it is true that they are corroding at different speeds and some may be dry or have electrically passive coatings or come from a reducing environment—recovery changes everything. Dry site recovered artifacts, left to dry even further, will be thrown artificially into a dry cycle. Acid inside the concretion will build up, concretions will begin to crack, and the artifact itself may begin to spall and flake. Even those objects that show no active corrosion and are kept in low relative humidity have acquired anions from their stay in the soil. These anion's such as Cl^- will initiate the corrosion process from within, perhaps years after recovery. This corrosion will show up in the form of weeping brown droplets that appear on the artifact. Concreted artifacts recovered from the ocean are also actively corroding and in need of stabilization. All corroded iron artifacts are surrounded by an acidic micro-environment, this micro-environment will encourage their continued destruction long after recovery.

Artifact stabilization begins, therefore, with recovery. Artifact storage for dry recovered iron, wet recovered iron, and oceanic iron is identical and begins at the site of recovery. All iron artifacts, concreted or not, should be wrapped in aluminum foil and placed in a basic solution directly after recovery (the only exception being composite artifacts—see Chapter 8). The foil should be tight but also allow the solution to get between it and the artifact.

This storage solution is no more than fresh water with the addition of a base to help neutralize the acid residing on the surface of the iron artifacts. Bases such as sodium carbonate or sodium bicarbonate (baking soda), are cheap, easy to find, safe, and serve a multipurpose as they can also be used during the electrolysis

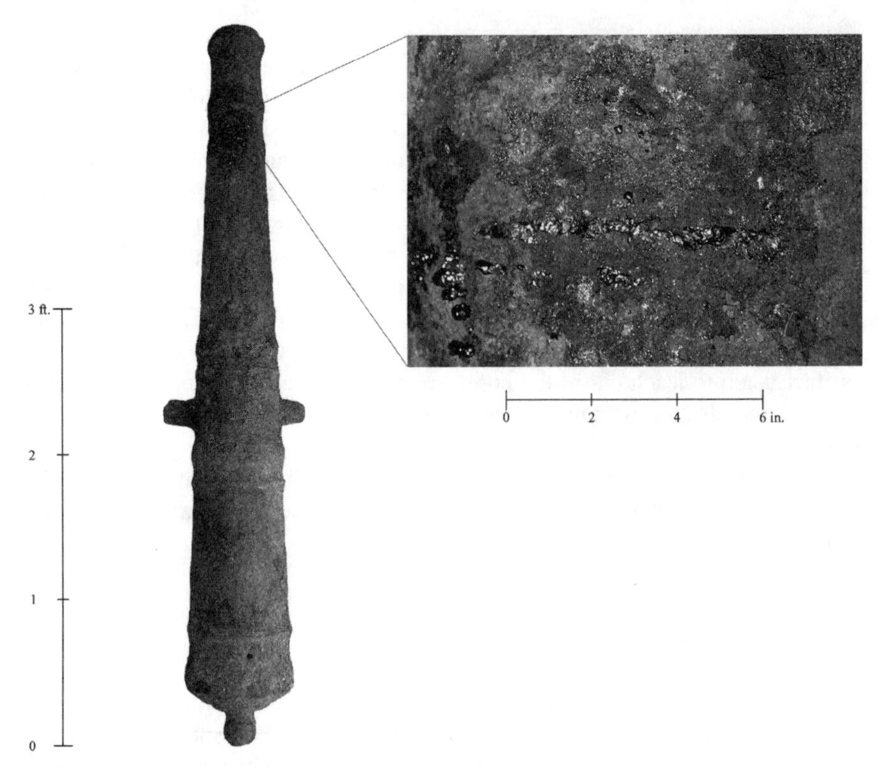

Figure 18. Weeping corrosion breaking out on 17th century cast iron cannon. The weeping appears as dark colored droplets which grow with an increase in relative humidity. Phtograph by Chris Valvano.

phase. Any basic solution will inhibit corrosion but storage solutions as high as 5% by weight of sodium carbonate (5gr. Sodium carbonate in 95 ml water) or the same with sodium bicarbonate, will produce a pH approaching 13, and will work nicely to help stabilize artifacts awaiting treatment.

The foil will actually begin the reduction of the artifact while it is undergoing storage on the basis of galvanic coupling. Since aluminum has a higher corrosion potential than iron it will donate electrons to the iron and reduce it (see section on Galvanic Wrap and Galvanic Coupling).

The storage solution should start with rainwater, deionized water, or distilled water. Tap water contains chloride anions that will encourage corrosion in iron. Since there are no worries concerning osmotic collapse (as there are in wood), there is no problem in placing iron artifacts directly into the alkaline storage water. Non-concreted artifacts and dry recovered artifacts, even those covered in earth concretions will almost immediately gain the benefits of the storage solution as it will easily penetrate to the artifact surface and neutralize much of the acid there while the foil will begin a beneficial ion exchange with the artifact.

Ocean formed non-porous concretions greatly complicate iron storage. It is doubtful that the basic storage solution penetrates to the well-protected acid layer at the artifact's surface and there is no reason to wrap these concretions in foil. The acid layer under the concretion has also become home to sulphate reducers and methanogens that can tolerate the anaerobic conditions and an acid 4.8 pH. Concretions formed on the sea floor and beneath the sea floor should be removed as soon as possible after artifact recovery or they will continue to promote corrosion in storage.

Artifacts fabricated of dissimilar metals should never be allowed to come into contact in the storage solution or they will form galvanic couples. In a galvanic couple the less noble metal will corrode to donate electrons to the more noble metal that will be reduced and protected. The storage solution will act as an electrolyte for galvanic couples and the corrosion rate will be extreme.

Iron artifact storage is only a temporary solution to the problem of artifact stabilization. The iron itself is porous and has absorbed large quantities of anions, these will have to be removed during processing to insure the continued stabilization of the object.

FRESH WATER RECOVERED IRON ARTIFACTS—METHODOLOGY

Iron objects recovered from fresh water lakes and rivers are sometimes found in pristine condition with no corrosion products and little or no concretion. It has been determined through experience that these artifacts need not undergo electrolytic reduction and are naturally stable. For treatment, these ferrous object are sent directly to the final wash stage.

SMALL NON-CONCRETED IRON ARTIFACTS—METHODOLOGY

Of all of the artifacts that will figuratively channel through the flow chart on page 69, this will be the smallest group. These are small fasteners, pins, decorative buckles, or broaches made of iron that seem to be in good condition and have little or no concretion. Above all they cannot have been recovered from a salt-water environment.

GALVANIC WRAP—METHODOLOGY

A galvanic wrap, known as electrochemical cleaning (Clarke and Blackshaw, 1982: 18), works on the principle of galvanic coupling. Since aluminum foil has a lower corrosion potential than iron it can be used as an anode to reduce the iron

Figure 19. Nearly perfect preservation exhibited by a 19th century flat iron recovered from a fresh water site. Photograph by Chris Valvano.

artifact. A galvanic wrap is a simple procedure. The artifact is simply wrapped tightly in aluminum foil and placed in an electrolyte contained in a beaker or glass—make sure the electrolyte gets between the foil and the artifact. The storage solution for iron (5% sodium carbonate or bicarbonate) will work well as the electrolyte. The time frame for this procedure is variable but results should appear within a few days and the entire procedure will be completed in a matter of a few weeks. If there is a question whether an artifact is destabilizing, the galvanic wrap can be placed on it indefinitely.

Results of the galvanic wrap should include a more defined surface, loss of any small amount of adhering carbonate, and a darkening of the surface of the artifact as it is reduced to magnetite. Magnetite is a stable non-reactive form of mineralized iron that frequently forms on the surface of reduced iron artifacts. All active ferrous oxy hydroxide (orange rust) should have disappeared during this process.

CONCRETION ANALYSIS—METHODOLOGY

Terrestrial brown mass concretions do not loan themselves well to analysis but there are simple procedures to test if the artifact is still intact inside the concretion. Hefting the artifact by hand will often reveal to a conservator whether there is doubt about the intact nature of an artifact. If the concretion feels too light to be made of iron then the metal has likely entirely corroded leaving only a mold. This concretion can be drilled and the hollow filled with epoxy resin to replicate the artifact. If there is doubt about metal residing inside a concretion a magnet can be placed on it. The magnet will only be attracted to metallic iron, not the oxides.

A common electrical multi-meter set for circuit continuity (ohms) and or resistance is a slightly more complex diagnostic tool used to test for artifact integrity. To effect this test the concretion is removed from two small places on the artifact and the probes of the multi-meter are touched to the artifact surface. If the meter registers a continuous circuit with little or no resistance, the artifact generally contains enough metal for conservation. It should be noted that in rare instances an artifact will contain enough metal for circuit continuity but not enough to survive electrolytic reduction, If there is any doubt the following radiographic testing should be done. Keep in mind that the tests listed above will succeed on 99% of the concretions that are tested, an X-ray machine is not considered necessary for the minimal intervention laboratory.

For more complex or important concretions, radiography should be contemplated. Medical radiographic machines are not usually powerful enough to penetrate iron contaminated concretion, if the conservator needs a radiograph they will probably need to go to an industrial facility with machines that can produce 120 to 350 kilovolts. X – ray photographs and flouroscopic images can often reveal detail of artifacts, such as their inner workings that cannot be gleaned in any other manner. If the artifact has deteriorated badly the x –ray may be the only data retrieved from it. Often a quick scan through an x – ray flouroscope will reveal whether an artifact is worth photographing. In a radiographic image the darker the shadow made by the artifact, the denser and easier it will be to conserve. Ironically, some iron artifacts can be in too good a condition to be molded in epoxy but at the same time too badly deteriorated for conservation using the recommended methods from this manual.

Concretions recovered from underwater sites can be examined and analyzed using virtually the same methods outlined above for terrestrial brown mass concretions. First they should simply be weighed by hand. Oceanic concretions can become much lighter than brown mass concretion as the iron in them can completely disappear, lost to the environment. In this case the cementite and calcium carbonate concretion will weigh only a fraction of what it should if iron is present. If there is doubt, however, a magnet and a multi-meter should suffice to complete the analysis. On rare occasions, when other artifacts are seen or suspected within the matrix of the concretion, a radiograph may be called for.

CONCRETION REMOVAL—METHODOLOGY

Generally speaking, concretion removal for dry recovered or oceanic artifacts depends on the robustness of the object as determined in the concretion analysis.

For solid heavy objects, a few hammer blows at a 90 degree working angle to the artifact surface will suffice to remove the majority of the concretion. Ocean formed concretion will tend to fall away completely whereas brown mass concretions will need more fine work at the laboratory. Much of the concretion can be removed in the field so long as the proper analysis is conducted ahead of time. Transportation of large quantities of concretion is unnecessary, expensive, and can lead to further corrosion of ocean recovered artifacts.

At the laboratory finer cleaning can be accomplished with the use of power tools such as an air or electric scribe (see Chapter 1). Dental picks and brushes compliment the air scribe as do frequent rinses under running non-chlorinated fresh water. This is an extremely messy procedure and when possible should be done outdoors.

If, for any reason, the concretions are allowed to dry during storage there will be several consequences. First, dry recovered artifacts may go into dry cycle break up. Oceanic concretions will absorb carbon from the atmosphere and become very hard. If there is a carbonized layer on the artifact it will adhere strongly to the concretion. On removal of the concretion this graphite layer (cast iron) will spall off taking the archaeological surface with it.

Only soft bristle brushes should be used to clean cast or pig iron artifacts after the concretion has been removed. A carbon matrix may remain on the surface of the artifact that looks like the original object but has no strength to resist abrasion.

After scrubbing rinsing and cleaning, the artifacts can receive a final cleaning with glass bead blasting. Blasting cabinets come in virtually any size depending on the intended use, industrially they are used to sand blast. In the laboratory less abrasive media is used, most often aluminum silicate beads. These beads will break on impact with an artifact absorbing much of the energy of the blasting (they must be replaced after some use). The glass beads come in fine, medium, and course sizes. Naturally the softer the surface to be worked the more care should be taken during the blasting phase. Cast or pig iron objects should not be subjected to more than 50 pounds of pressure per square inch in the blasting unit. Fine and medium grit blasting media should also be used. Usually the unit should be initiated at about 25 pounds pressure and adjusted upward if the blasting proves ineffective. The nozzle should be held four to six inches from the subject.

Wrought iron can easily withstand 50 to 100 pounds of pressure and a medium to course grain size at the same nozzle distance to the subject as cast iron. Small areas on any object should be blasted and observed before the entire artifact is cleaned.

The blasting pressure for steel depends entirely on the condition of the metal and can range from 25 to 100 pounds. Naturally the more robust the metal the higher the pressure than can be exerted.

A blasting cabinet is not part of the original set up of a minimal intervention laboratory. However, if metals are recovered and conserved on a regular basis, it and an air scribe should be first priority in equipment purchased after lab set-up.

LOW AMPERAGE ELECTROLYSIS—METHODOLOGY

Low amperage electrolytic reduction or electrolysis should be used on cast and pig iron to reduce the risk of exfoliating any outer carbonized layer that may be present. During the process of electrolysis, hydrogen gas is given off by the artifact at the metal surface. If the electrical amperage is limited the gas production is minimal and so are the chances that hydrogen production will loosen or destroy the carbonized outer layer.

Electrolysis is an electrochemical process whereby the corrosion product accumulated on the surface layer of an artifact is reduced to a more stable form of iron mineral. This mineralized corrosion product has a smaller surface area than common rust allowing for an increase in the artifact's surface porosity. Chloride ions that have penetrated the metal are then more easily rinsed out of the metal and into the electrolytic solution.

Electrolytic set up is not complex. First a cleaned artifact is placed in one of the tanks used for this purpose. Be sure the exhaust fan in the lab is turned on as the hydrogen and oxygen gases given off during the process can be a very explosive combination.

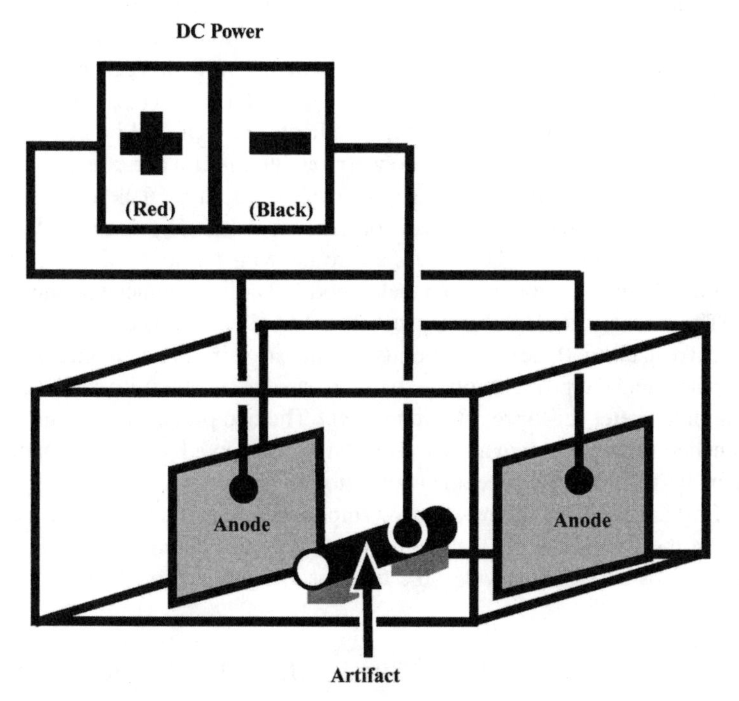

Figure 20. Set up for electrolytic reduction. The artifact should not be allowed to touch the anodes but the anodes should closely surround the artifact. Illustration by Nathan Richards.

Next the electrolytic solution is prepared. Enough water should be added to the tank to more than cover the artifact. Electrolytic solution is made by mixing a .25%–.5% sodium carbonate (Na_2CO_3soda ash), or sodium bicarbonate ($NaHCO_3$ – baking soda) solution. This calculates to .25 grams to .5 grams of soda per 99.75–99.5 mls of water (100mls is close enough). Solution percentage is used to control the amperage or amount of electricity flowing through the system. The more sodium carbonate or bicarbonate added to the solution the more electricity can be carried by the free ions, hence a higher amperage will be registered for a .5% solution than a .25% solution. Amperage registered on the battery charger with light electrolyte solutions of .25% to .5% should range from less than 1 amp to 7 amps depending on the surface area of the artifact (see note on light concentration electrolyte).

After the elelctrolyte is prepared and placed in the tank two mild steel anodes are placed adjacent to the artifact. Be sure the anodes do not touch the object as this will create a dead short. The sacrificial anodes should be connected to the positive terminal (RED) of the DC power source with alligator clips. Charging both anodes can be done by clamping the DC power source clip to one of the sacrificial anodes and cross connecting the two anode plates with a small cross connect alligator wire with clips. Locating the alligator clips on top of the anodes, out of the electrolytic solution will save having to replace them. In solution they will corrode readily. All connections should be made to bare shiny metal on the sacrificial anodes, alligator clips placed on painted or corroded surfaces will not conduct current. A file or sandpaper can be used to create a bare shiny contact point on each anode.

The artifact is then connected to the negative terminal (BLACK) of the DC power Source. This can be done with an alligator clip (it will be protected and reduced with the artifact so there is no worry about this clip corroding). The point of contact must be bare shiny metal, so an out of the way area on the artifact must be cleaned through concretion and corrosion product and the clip connected to it. An alternative procedure would be to strip a wire and wrap the stripped section around the artifact so long as it touches and makes good electrical contact with the artifact.

The most frequent problem with electrolysis is that the contact points offer electrical resistance. If the contact points are not bare shiny metal the circuit will not be complete and the process will not work. Connection continuity can be checked with a multi-meter set for resistance in ohms. The two probes on the multi-meter are touched to the upstream and down-steam side of an alligator clip connection, if the meter reads 0 resistance, the connection is continuous.

Cast and pig iron will receive low amperage electrolysis at 6 volts. The amp meter on the DC power source should move between 0 and 4 amps. In a short period the artifact will be giving off a fine mist of hydrogen bubbles.

HIGHER AMPERAGE ELECTROLYSIS—METHODOLOGY

Higher amperage electrolytic reduction is reserved for wrought iron artifacts. Unlike cast and pig iron, there is little danger of damaging wrought iron with too much current and resulting gas generation. Normally gas generation should be kept to a minimum as it will interfere with the chloride rinse, but in certain instances it can actually help in cleaning soil from cracks and fissures and to remove stubborn concretions.

Higher amperage electrolysis is set up the same as low amperage electrolysis. The only difference in higher amperage electrolysis is that a higher percentage sodium carbonate or bicarbonate solution is prepared (.5% to 1%) and the power source is switch to 12 volts rather than 6. Most power sources cannot sustain a production of more than 10 amps without automatically shutting down, so too high a percentage of sodium carbonate or bicarbonate will carry too much load and eventually shut down the battery charger. Ten amp power production at 12 volts can be felt as a tingling sensation, should you place your hand in the electrolyte, but there is no danger whatever of electrocution.

NOTE ON LIGHT CONCENTRATION ELECTROLYTE

The minimal intervention laboratory uses a very dilute electrolyte solution for several reasons. First the direct current (DC) power sources used in a minimal intervention laboratory are battery chargers that can be purchased from any hardware or automotive store. They have no amperage controls and will basically supply as much electricity as they are rated for (example 6 or 10 amps). The dilute electrolyte solution keeps the current load within the capacity of the charger (1–6 or 10 amps) and represents an extremely low current density on the artifact. Minimal hydrogen gas production is the observable goal in electrolysis. Overt gas production simply interferes with the electrolytic rinse. This minimal gas production can be best achieved at anywhere from 50 milliamps to 2 amps depending on the size of the artifact (the greater the surface area the greater the amperage).

Battery chargers as DC power sources are cost affective, affordable, and easy to use (simply place on manual and select 6 volts). Experience has demonstrated that there is *no* danger from power surges producing enough gas to exfoliate the outer layers of an object. Should this still be a concern, the chargers are easily modified with a capacitor (in the 3300 micro ferrid range) wired between the positive and negative outflow if it is desired to have a smoother current supply.

The minute amount of soda in the electrolyte solves several other problems reported by laboratories. The alkaline soda solutions are so dilute and non-toxic that disposal is never a problem, in fact the solution can easily be ingested or splashed in the eyes with no ill effects. Some other caustic electrolytes such as sodium hydroxide (NaOH) can actually burn the skin on contact with an artifact after weeks of rinsing. Electrolysis with these dilute sodium carbonate or bicarbonate solutions can be undertaken on artifacts that are wood and iron composites (see chapter 8), as the electrolyte does not harm most wood in the time frame given for the treatment.

Some laboratories report insoluble carbonate precipitates on their artifacts when using high concentrations of soda as an electrolyte. High concentrations are unnecessary and actually counter productive as they must be neutralized for disposal and will carry too large an electrical load to keep gas production to a minimum.

Finally it appears that most laboratories use a high concentration electrolyte to reduce the amount of corrosion on the sacrificial anodes. This is understandable if the treatment tanks are made of mild steel and the tank itself is used as an anode. The anodes used at a minimal intervention lab are scrap metal placed in a tank. These cheap and easily obtained anodes are sacrificed to protect both the conservator and the artifact, in this light the anodes cost is negligible. It should be noted that as the anode corrodes, the treatment water will become cloudy with foam on the surface while it turns red from the corrosion products, this is to be expected. The electrolyte will settle and become clear again in a matter of a few days.

CHLORIDE TESTING—METHODOLOGY

Since the principle of electrolysis is based on reducing corrosion products and rinsing anions from the iron, the amount of anions, particularly chlorides, in the electrolytic solution should give a good indication of how far along the rinse is. Though this is true in theory there has never been direct analytical correlations established between the onset of reactivated corrosion in an artifact and the amount of chlorides rinsed during treatment, or between the amount of chlorides present in the rinse solution and those left in the metal.

Other problems exist with using chloride content as an indicator of treatment duration. The measurable chloride content of a small artifact placed in a large treatment tank will not be sufficient to change the overall chloride content of the electrolyte. This may trick an uninformed conservator into concluding there were no chlorides in the artifact. Therefore, it should be understood, that chloride testing is not by any means an absolute measure, nor is it essential to doing electrolysis. Chloride testing is dependant on the ratio of artifact surface area to treatment tank volume, the amount of chlorides present in the artifact, the size of the artifact in relation to the tank volume, the porosity and material type of the artifact. For the purpose of conservation in this manual, chloride testing and graphing is only recommended for large artifacts, those that will undergo months or years of treatment.

In principle chloride testing works as a graph of chlorides measured in parts per million (ppm) in the rinse solution plotted against time. When the chlorides rise to a certain amount in the rinse solution it is changed so a back-pressure of chlorides will not build up and inhibit further rinsing (see chart).

Generally speaking when the weekly test of chlorides reveals a level of 1000 to 1500 ppm or more, the solution is changed and the anodes are scrubbed free of corrosion. It is also a good idea at that time to scrub and rinse the artifact with nylon brushes to remove stubborn concretion and free loose surface debris and corrosion

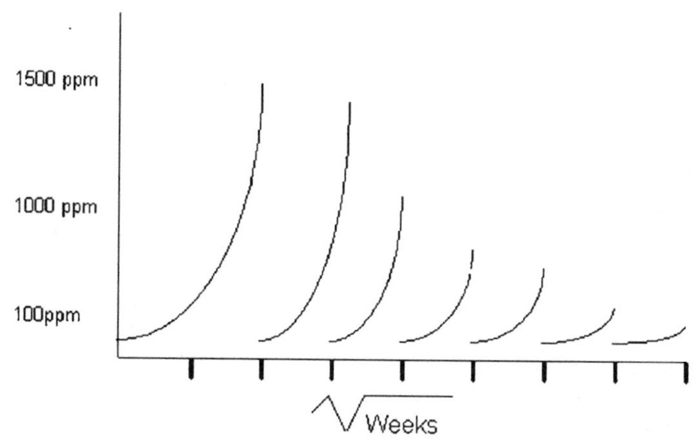

Figure 21. Graph of chloride rinsing in a large iron artifact in parts per million of chloride ion concentration vs. time in weeks squared (64 weeks total).

products. After treatment has proceeded for several weeks a concentration of 500 ppm or more will signal the need for cleaning and change of solution. Eventually the solution will be changed at 100 ppm and distilled, rain, or deionized water can be substituted for tap water to eliminate any external source of chlorides from the process. When the chloride ion concentration levels at something near 50 ppm the treatment is discontinued and the artifact is ready for the final washing.

There are many ways to test for chloride ions, some of which are themselves a source of toxic chemical production. The actual test presented in this manual is simple, safe, and non-polluting. It can be accomplished by placing a small sample of the electrolyte in a clean 50 ml beaker. A chloride titration strip (available at scientific suppliers) is placed into the beaker. When enough solution wicks to the top of the strip the white indicator line on the strip is read and compared to the chart for ppm chloride ions. If the electrolytic reduction is initiated in tap water the city-normal chloride ion concentration, should be subtracted from the total chloride reading to compute an accurate chloride rinse reading. Larger artifacts such as cannon and anchors will take up to 4 years of electrolytic reduction.

Simple time frames serve as good references in the treatment times for smaller artifacts. Pin, nail, or buckle sized iron artifacts should receive one to three weeks of electrolysis. The solution should be changed weekly only if the artifacts are heavily contaminated with salt and they are in a ratio of one artifact per liter of solution or higher. Larger artifacts such as iron drift bolts, cast machine parts, spikes, or rifle barrels should undergo up to three months of treatment, with bimonthly changes of solution.

In the end, the only indisputable method of determining if the electrolysis treatment time has stabilized the artifact is to observe it in storage. If, at any time (perhaps years later) it begins to corrode or weep an oily brown liquid, the artifact needs further reduction treatment.

FINAL WASH—METHODOLOGY

When an artifact has completed electrolysis it should undergo a final scrub with nylon bristle brushes and a paste made from sodium bicarbonate and distilled or deionized water. A rinse and hour long soak in distilled water will insure that most of the electrolyte and bicarbonate have been removed.

Boiling a robust artifact for several hours in a 2% sodium carbonate, or bicarbonate solution, can help deep clean an artifact. This should not be done if the artifact has any fragile or friable areas. Afterward the object should be rinsed and soaked in distilled water. At this point all artifacts have been cleaned of chlorides and should not be handled without gloves for the remainder of the treatments.

DEHYDRATION—METHODOLOGY

There are two basic types of dehydration recommended for iron, oven drying for wrought iron and solvent drying for cast. Wrought iron is more robust than cast and has no surface carbonized layer that could flake off during the dehydration process. Additionally, though wrought iron will corrode along the silicate slag intrusions there will, in all probability, be no fissures that penetrate to the interior of the artifact, as there will be in some cast iron objects. These fissures could be pressurized, should the water in them turn to steam, and cause the artifact to break apart.

Oven dehydration for wrought iron artifact begins with the oven preheated to 350 degrees F, or 177 degrees C. Remove the artifact from the distilled water bath and place it in the oven for 24 to 48 hours. The length of dehydration depends on the surface texture of the artifact. Generally, the rougher the texture the longer the dehydration period. Most often with this method a small amount of orange iron oxide will form before all of the water evaporates. This oxide is no problem, and can be cleaned with a cloth dipped in alcohol, in any case the presence of the oxide will not effect the final esthetic appearance of the artifact.

Solvent dehydration is a much more delicate process and can be used on cast and pig iron artifacts as well as all others that may be heat sensitive such as soldered cans or historic pieces that may still have paint or surface treatments. Solvent drying subjects an artifact to three successive baths in acetone or denatured alcohol. Organic solvents are extremely flammable and great care should be taken they are kept from any ignition source. They may also pose other health risks and should be handled without inhaling the vapors or prolonged contact with the skin.

Each bath should take at lest an hour. When the artifact is removed from the first bath it should be placed into fresh solvent and the process repeated to insure complete water removal. This is obviously much more expensive and dangerous than oven drying but it is gentler to the objects being treated. Solvent tubs should be covered and clearly marked. Used solvent should be stored in approved metal cans and can be used to treat more than one artifact.

TANNIC ACID APPLICATION—METHODOLOGY

According to the Material Safety Data Sheets, tannic acid can be a carcinogen with prolonged exposure. Its use should be restricted to outdoors or well ventilated rooms. It should only be handled with rubber gloves and a breathing mask should be worn during treatment. Also its application to the surface of an artifact is not mandatory. Esthetics and or archaeological reasons may dictate that it not be used.

Tannic acid is applied to the exterior of an artifact for two reasons. First, although it is an acid, it is a corrosion inhibitor. This acid forms complex organic compounds that help stabilize the surface of the artifact being treated. The second reason to use tannic acid is that it restores the artifact to a more historically correct black color.

To apply the tannic acid the artifact is allowed to cool from oven drying, or air dry from solvent drying. A 5% mixture of tannic acid in alcohol (5 grams tannic acid to 95 ml. alcohol) is brushed on the surface of the artifact. Several coats may be necessary to thoroughly darken the object. Water should not be used as the solvent medium for the tannic acid for the obvious reason that the artifacts were just dehydrated.

PROTECTANT APPLICATION—METHODOLOGY

A final humidity barrier should be applied to the surface of most iron artifacts to prevent moisture from reentering the object and promoting corrosion. This coating should be applied immediately after the tannic acid application or dehydration if no acid was applied.

Small to medium sized artifacts can be safely coated in microcrystalline wax (micro-wax can be obtained from conservation suppliers). The wax application pan should be a high sided large container (like a turkey roaster) that can fit on the stove top and be serviced by more than one burner. Micro-wax has a high melting temperature for wax at about 175 degrees F (79 degrees C). The micro-wax is melted on the stovetop with burners set to medium. The exhaust fan above the stove should be turned on as the fumes from the melting wax can be irritating. Once the wax is melted the burners can be turned between medium and low the wax should remain at about 210 to 220 degrees F (98.9 to 104 degrees C). A wire basket does well to support most artifacts while undergoing treatment and saves the conservator from having to fish in the wax with a utensil to retrieve the artifact.

When the artifact is placed into the melted wax it will begin to bubble vigorously as more water is driven from the objects surface by the heat of the wax. The bubbling usually subsides in a couple of hours or less. When it has stopped turn the stove top burners to low and wait one hour to remove the wire basket and artifact from the wax. Place the basket on a pan to drip excess wax while it cools. Excess wax can be removed with a knife after the artifact has completely cooled or smoothed with the fingers before the wax has totally solidified.

There are a few dangers involved in micro-wax coating artifacts. Danger of burns is ever present around melted wax, great care should be taken to prevent this and heavy thermal gloves are recommended when working around the stove. Fire is another danger when dealing with hot wax. To prevent this the micro-wax should never be left on without a technician in the room.

The micro-wax coating both dehydrates and protects the artifacts from moisture in the atmosphere. It can be reapplied at any time or removed in boiling water or an oven. Unfortunately micro-wax is impractical to use it on larger artifacts.

Shellac is another coating that has been used successfully for years by conservation technicians as a moisture barrier. It is easily applied with a paint brush or spray applied. Shellac can be thinned or removed with alcohol should it become necessary to do so.

Finished iron artifacts are safe to handle and should be stored under climatically controlled conditions. Ideally the temperature should not vary more than a few degrees and the relative humidity range should be kept around 40% to 50%.

CONCLUSION

Iron artifacts represent a true challenge to the archaeologist/conservator. Of the many processes developed for the treatment of iron, only a few truly fit into the safe, non-toxic, and easy to use category. Low amperage, light load electrolysis, and galvanic wrapping are safe, effective, and easy. The results of conserving all the iron artifacts recovered from an archaeological site should be dramatic. Interpretation will be easier and more accurate. Micro-excavation will reveal detail not seen before and a stabilized artifact can be revisited even after indefinite time periods in storage.

Conservation of iron and other metals truly begins with adequate planning before the retrieval of iron artifacts on an archaeological project, land or underwater. As will be seen, all metal artifacts will benefit from stabilization recovery in which the artifacts are place in an alkaline storage solution and in most cases wrapped in aluminum foil for transportation to the lab.

CHAPTER 3: ARCHAEOLOGICAL IRON

Adams, Christopher. "The Treatment of the BMW 801D-2 Radial Aero Engine Rescued from the Loiret River." The Laboratories of Groupe Valectra: Electricite de France, 1992. [ECU-581]

Aldaz, A., T. Espana, V. Montiel, and M. Lopez-Segura. "A Simple Tool for the Electrolytic Restoration of Archaeological Metallic Objects with Localized Corrosion." *Studies in Conservation* (1986): 175–176. [ECU-0386]

Allsopp, Dennis & Kenneth J. Seal. "An Introduction to Biodeterioration: Metals." *An Introduction to Biodeterioration*. Edward Arnold, London 1986. [ECU-570]

"All Right Troops—Fan Out and Find Every Last Artwork." *Smithsonian* Vol. 26 No. 1, 1995, pp. 140–149. [ECU-783]

Alsop, Joseph. "Warriors from a Watery Grave." *National Geographic* 163 (1983): 820–827. [ECU-271]

Ambrose, W.R. and M.J. Murrinery. "Secondhand Metal Conservation and Industrial Archaeology of Shipwreck Artifacts." *Archaeometry: Further Australasian Studies* (Canberra, Australia) (1987): 280–291.

Anon. "Facts Surface about Bugs and Disappearing Oil Rigs," *New Scientist* 87 (1980): 591. [ECU-0277]

Anon. "Making a Meal of Iron," *Science News* 132 (1987): 104. [ECU-106]

Anon. "Suggested Techniques and Equipment for Cleaning and Stabilizing of Corroded Iron Artifacts." [ECU-572]

Argo, James. "A Qualitative Test for Iron Corrosion Products." *Studies in Conservation* 26, No. 4 (1981): 140–142. [ECU-166]

—. "On the Nature of Ferrous Corrosion Products on Marine Iron." *Studies in Conservation* 26 (1981): 42–4. [ECU-120]

—. "The Treatment of Corrosion with Amines." *Conservation News* 17 (1982): 7–9.

Arrhenius, O. "Corrosion on the Warship *Wasa*." (Swedish Corrosion Institute Bulletin) 48 (1967).

Ashton, John. "Conservation Notes: Maintenance of Japanese Swords." Australian War Memorial. [ECU-583]

Ashworth, V. and R.P.M. Procter. "The Role of Coatings in Corrosion Prevention—Future Trends." *Journal of the Oil and Colour Chemistry Association* 56 (1973): 478–490. [ECU-133]

Baker, H.R., R.N. Bolster, P.H. Leach, and C.R. Singleterry. "Examination of the Corrosion and Salt Contamination of Structural Metal from the USS *Tecumseh*." (Naval Research Laboratory Memorandum) (1987).

Barker, B.D., K. Kendell, and C. O'Shea. "The Hydrogen Reduction Process for the Conservation of Ferrous Objects." *Conservation of Iron* (1982): 23–27. [ECU-279]

Barkman, L. "Preservation of Large Iron Objects." *Proceedings* (First Southern Hemisphere Conference on Marine Archaeology) (1987): 127–128.

Biasini, V. and E. Cristoferi. "A Study of Corrosion Products on Sixteenth and Seventeenth Century Armour from the Ravenna National Museum." *Studies in Conservation* Vol. 40 No. 4, 1995, pp. 250–256. [ECU-660]

Birchenal, L.E., and R.A. Meussuer. "Principles of Gaseous Reduction of Corrosion Products." *NBS Special Publication* (United States) (1977): 3857.

Black, J.W.B. "Choosing a Conservation Method for Iron Objects." *Conservation of Iron* (1982): 15. [ECU-281]

Blackshaw, S.M. "An Appraisal of Cleaning Methods for Use on Corroded Iron Antiquities." *Conservation of Iron* (1982): 16–22. [ECU-280]

Bobichon, C., C. Degrigny, F. Dalard, and Q.K. Tran. "An Electrochemical Study of Iron Corrosion Inhibitors in Aqueous Polyethylene Glycol Solutions." *Studies in Conservation*. Vol. 45 No. 3, 2000, pp. 145–153. [ECU-661]

Bradley, G.W. and J.A. Smith. "Removal of Iron Oxide Deposits without Feric Ion Corrosion." *Materials Protection and Performance* 12 (1973): 48–52. [ECU-130]

Brady, P.R., G.N. Freeland, R.J. Hine, and R.M. Hoskinson. "The Absorption of Certain Metal Ions by Wool Fibers." *Textile Research Journal* 44 (1979): 733–735.

Bright, Leslie S. "Electrolytic Reduction of Two Large Cannon from the USS Peterhoff." (Unpublished Manuscript, North Carolina Department of Archives and History) [n.d.]. [ECU-423]

Bright, Leslie S. "Preservation of a Cannon from the USS *Peterhoff*." (Unpublished Manuscript, North Carolina Department of Archives and History) (1967). [ECU-422].

Bright, Leslie S. "Recovery and Restoration of Steam Machinery from the Snagboat *General H.G. Wright*." *Proceedings* (Underwater Archaeology, Society for Historical Archaeology Conference) (1989): 121–143. [ECU-17]

Brown, B.F., H.C. Burnett, et al., ed's. *Corrosion and Metal Artifacts: A Dialogue Between Conservators and Archaeologists and Corosion Scientists* (U.S. Department of Commerce, National Bureau of Standards Special Publication No. 479, Washington D.C.) (1977).

Brownsword, R. and E.E.H. Pitt. "A Medieval Weight from the *Mary Rose*." *International Journal of Nautical Archaeology* 15 (1986): 161–163. [ECU-157]

Burgess, David. "Metals and their Conservation." *Chemical Science and Conservation*. London: Macmillan, 1990. [ECU-575]

Butler, G. and H.C.K. Ison. *Corrosion and its Prevention in Waters*. New York: Robert E. Krieger, 1978.

Campbell, Harry H. *The Manufacture and Properties of Iron and Steel*. New York: McGraw Hill, 1907.

Campbell, H.S. "The Corrosion Factor in Marine Environments." *Metals and Materials* I:8 (1985): 479–483. [ECU-135]

Campbell, H.S. and D.J. Mills. "A Metallurgical Study of Objects Salvaged from the Sea." (Report CDC, BNF Metal Users Consulting Service) (December 1976).

—. "Marine Treasure Trove—A Metallurgical Examination." *Metallurgist and Materials Technologist* 9/10 (1977): 551–556. [ECU-0224]

Cantelas, Frank. "Aircraft Corrosion." E-mail sent to Brad Rodgers, December 2001, Greenville, NC. [ECU-585]

Carlin, Worth and Donald H. Keith. "An Improved Tannin-Based Corrosion Inhibitor-Coating System for Ferrous Artifacts." *International Journal of Nautical Archaeology*. Vol. 25 No. 1, 1996, pp. 38–45. [ECU-663]

Carline, Worth, Donald Keith, and Juan Rodriguez. "Galvanic Removal of Metallic Wrought Iron from Marine Encrustations." *International Journal of Nautical Archaeology* Vol. 31 No. 2, 2002, pp. 293–299. [ECU-748]

CCI Laboratory Staff. "An Inexpensive and Simple Tacking Iron." *CCI Notes* (Canadian Conservation Institute) 18/1 (1995). [ECU-551]

CCI Laboratory Staff. "Care and Cleaning of Iron" *CCI Notes* (Canadian Conservation Institute) 9/6 (1995). [ECU-537]

CCI Laboratory Staff. "Storage of Metals" *CCI Notes (*Canadian Conservation Institute) 9/2 (1995). [ECU-536]

CCI Laboratory Staff. "Tannic Acid Treatment." *CCI Notes* (Canadian Conservation Institute) 9/5 (1989). [ECU-465]

CCI Laboratory Staff. "Recognizing Active Corrosion." *CCI Notes* (Canadian Conservation Institute) 9/1 (1991). [ECU-354]

Cheek, P. "Conservation of Cannon at Chatham Historic Dockyard." *International Journal of Nautical Archaeology* 15 (1986): 253–254. [ECU-156]

Clarke, R.W., and Susan M. Blackshaw, eds. *Conservation of Iron*. (Monograph No. 53, National Maritime Museum, Greenwich, U.K.) (1982). [ECU-289]

—. *Conservation of Iron* (Monograph No. 53, National Maritime Museum, Greenwich, U.K.) (1982). [ECU-289]

Corfield, Michael. "Radiography of Archaeological Ironwork." *Conservation of Iron* (1982): 8–14. [ECU-282]

"Corrosion on Riotinto Pier at Huelva." *British Corrosion Journal* Vol. 28 No. 1, 1993, pp. 253–258. [ECU-664]

"Corrosion in Saline and Sea Water." *Corrosioni Prevention and Control.* Vol. 39 No. 4, 1992, pp. 99–104. [ECU-665]

"Cor-trol Vapor Corrosion Inhibitors." Corroless North America, Stamford, CT. [ECU-584]

Costerson, J. William, et al. "Bacterial Biofilm in Nature and Disease." *Annual Review of Microbiology* 41 (1987): 435–464.

Dalard, F., Y. Gourbeyre, and C. Degrigny. "Chloride Removal from Archaeological Cast Iron by Pulsating Current." *Studies in Conservation* Vol. 47 No. 2, 2002, pp. 117–121. [ECU-666]

Daniels, L., N. Belay, B.S. Rajagopal, and Paul J. Weimer. "Bacterial Methanogenesis and the Growth for CO_2, with Elemental Iron as the Sole Source of Electrons." *Science* 237 (1987): 509–511. [ECU-26]

Dettner, Heinz W. *Elsevier's Dictionary of Metal Finishing and Corrosion* (1971).

Dillmann, P., R. Balasubramaniam, and G. Beranger. "Characteristics of Protective Rust on Ancient Indian Iron Using Microprobe Analysis." *Corrosion Sciencel* Vol. 44 No. 10, 2002. [ECU-662]

Duncan, S.I. "Chloride Removal from Iron: New Techniques." *Conservation News* 31 (1986): 21–23.

Dunton, John V.N. "The Conservation of Excavated Metals in the Small Laboratory." *The Florida Anthropologist* 7:2 (1964): 37–43. [ECU-421]

Edwards, Marc and Nicolle Sprague. "Organic Matter and Copper Corrosion By-Product Release: A Mechanistic Study." *Corrosion Science* Vol. 43 No. 1, 2001. [ECU-667]

Eriksen, Egon and Svend Thegel. "Conservation of Iron Recovered from the Sea." *Tojhusmuseets Skrifter* (1966): 8.

—. "Conservation of Iron Recovered from the Sea." *Tojhusmuseets Skrifter* (Abstract and Conclusion) (1966): 93–97. [ECU-0030]

Evans, Ulick R. "The Rusting of Iron: Causes and Control." *Studies in Conservation* (London) No. 7 (1972).

"Facts Surface about Bugs and Disappearing Oil Rigs." *New Scientist* 87 (1980): 591. [ECU-277]

"F6F Hellcat Preservation Plan." Revision: A 950424. [ECU-589]

Fem, Julia D., and Kate Foley. "Passivation of Iron." *Conservation in Archaeology and the Applied Arts* (Contributions to the 1975 Stockholm Conference, N. Brommelie and P. Smith, eds., IIC-London) (1975): 195–198.

Foley, V.P. "Another Method for the Treatment of Ferrous Artifacts." *Florida Anthropologist* 18:3 (1965):65–68. [ECU-396].

—. "Suggested Design and Construction for Small Laboratory Electrolysis Apparatus." *Proceedings* (Conference on Historic Site Archaeology Papers 1965–1966, Vol. 1) (1967): 100–110.

Ford, Bruce. "Materials for the Traditional Care of Japanese Swords." [ECU-592]

Forth, Karl D. "Detox for Airplanes." *Aviation Equipment Maintenance* 10:8 (1991): 46–47. [ECU-365]

Friendly, Alfred. "An Ocean Relic is Given New Life." *Smithsonian* 5:12 (1975): 90–95. [ECU-275]

Fontana, Bernard L. "The Tale of a Nail: on the Ethnological Interpretation of Historic Artifacts." *The Florida Anthropologist* Papers of the 5th Annual Historic Sites Conference. Vol. 18 No. 3, Part 2, September 1965.

Geckle, Robert A. "How Ultrasonics Gets the Dirt Out." *Metal Progress* 108 (1975): 35–39. [ECU-225]

Gilburg, Mark R. "Storage of Archaeological Iron in Deoxygenated Aqueous Solutions." *Journal of the IIC-CG* 12 (1987): 20–27. [ECU-123]

Gilberg, Mark R., and Nigel J. Seeley. "The Identity of Compounds Containing Chloride Ions in Marine Iron Corrosion Products: A Critical Review." *Studies in Conservation* 26 (1981): 50–56. [ECU-223]

—. "The Alkaline Sulphite Reduction Process for Archaeological Iron: A Closer Look." *Studies in Conservation* 27:2 (1982): 180–184. [ECU-123]

—. "Liquid Ammonia as a Solvent and Reagent in Conservation." *Studies in Conservation* 27 (1982): 38–44. [BCU-395]

Green, Jeremy N., G. Henderson, and N. North. "A Carronade from the *Brig James*: Its History, Conservation, and Gun Carriage Reconstruction." *International Journal of Nautical Archaeology* [n.d.] [ECU-244]

Greene, Virginia. "The Use of Benzotriazole in Conservation." (ICOM Committee for Conservation, Madrid) (1972).

—. "The Use of Benzotriazole in Conservation." *ICOM Proceedings* (Committee for Conservation, 4th Triennial Meeting, Venice) (1975): 1–10.

Gregory, David. "Monitoring the Effect of Sacrificial Anodes on the Large Iron Artifacts on the Duart Point Wreck, 1997." *International Journal of Nautical Archaeology* Vol. 28 No. 2, 1999, pp 164–173. [ECU-670]

Hamilton, Donny L. *Conservation of Metal Objects from Underwater Sites: A Study in Methods*. Austin: Texas Memorial Museum & The Texas Antiquities Committee, 1976. [ECU-4]

—. "Electrolytic Cleaning of Metal Articles Recovered from the Sea." *Science Diving International* (N.C. Flemming, ed., Proceedings of the Third Symposium of the Scientific Committee of the Confederation Mondiale des Activities Subaquatiques, London) (1973): 96–104.

—. *Basic Methods of Conserving Underwater Archaeological Material Culture*. Washington, D.C.: U.S. Department of Defense Legacy Resource Management Program, January 1996. [ECU-610]

Hansen-Hjelm, Nils. "Cleaning and Stabilization of Sulphide-Corroded Bronzes." *Studies in Conservation* 29 (1986). [ECU-0241]

Hodges, Henry. "Iron and Steel." *Artifacts*. London: John Baker, 1964. [ECU-571]

Hoyt, Cathryn. "Protecting Damaged Concretions Underwater." (Unpublished Manuscript, Bermuda Maritime Museum) (1986). [ECU-485]

"Inhibition of Corrosion of Iron and Aluminum." *British Corrosion Journal* Vol. 31 No. 3, 1996, pp. 161–240. [ECU-671]

Jobling, Jim. "The Conservation of Cast Iron Cannon from the Chandeleur Islands." *Proceedings* (Underwater Archaeology, Society for Historical Archaeology, Tucson, Arizona) Toni Carrell, ed. (1990): 97–100.

Katzev, Michael J. and F.H. Van Doominck, Jr. "Replicas of Iron Tools from a Byzantine Shipwreck." *Studies in Conservation* 2:3 (1966): 133–142. [ECU-343]

Keene, Suzanne, ed. *Corrosion Inhibitors in Conservation*. (Occasional Papers No. 4. United Kingdom Institute for Conservation, London) (1985).

Keene, Suzanne, and Clive Orton. "Stability of Treated Archaeological Iron: An Assessment." *Studies in Conservation* 30 (1985): 136–142. [ECU-494]

Knight, B. "Why do Some Iron Objects Break Up in Store?" *Conservation of Iron* (1982): 50–55. [ECU-286]

Knowles, E. and T. White. "The Protection of Metals with Tannins." *Journal of Oil and Colour Chemists Association* 41 (1958): 10–23. [ECU-495]

LaQue, F.L. "Corrosion by Sea Water: Behaviour of Metals and Alloys in Sea Water." *The Corrosion Handbook* H.H. Uhlig, ed., New York: John Wiley and Sons, 1975: 383–430.

Lemer, G.M. "The Cleaning and Protective Coating of Ferrous Metals." *Bulletin of the IIC-AG* 12:2 (1972): 97–108.

Lewis, W.D. *Iron and Steel in America*. Hagley Museum, Wilmington, Delaware, 1976.

Logan, Judith A. "Conservation of Corroded Iron Artifacts—New Methods for On-site Preservation and Cryogenic Deconcreting." *International Journal of Nautical Archaeology* 16:1 (1967): 49–56. [ECU-153]

—. "An Approach to Handling Large Quantities of Archaeological Iron." *Proceedings* (ICOM Comittee for Conservation, 7th Triennial Meeting, Copenhagen) (1984). [ECU-360]

—. "The Longest Treatment in the History of CCI," *CCI Newsletter* Spring–Summer (1989): 4–6. [ECU-50]

—. "Tannic Acid Treatment." *CCI Notes* 9/5 (1991). (ECU-351]

Lucey, V.F. "The Mechanism of Dezincification and the Effect of Arsenic, I." *British Corrosion Journal* I (1965): 9–14. [ECU-141]

MacLeod, Ian Donald. "A Microphotographic Review of Corrosion Phenomena from a Shipwreck." *ICOM Bulletin* 9 (1981): 92–96.

—. "In-Situ Corrosion Studies on the Duart Point Wreck, 1994" *International Journal of Nautical Archaeology* Vol. 24 No. 1, 1995, pp. 53–59. [ECU-672]

—. "The Application of Corrosion Science to the Management of Maritime Archaeological Sites." *Bulletin of the Australian Institute for Maritime Archaeology* 13:2 (1989): 7–16.

—. "The Electrochemistry and Conservation of Iron in Seawater." *Chemistry in Australia* 56:7 (1989): 227–229.

—. "The Longest Treatment in the History of CCI." *CCI Newsletter* (Spring–Summer 1989): 4–6. [ECU-50]

—. "Conservation Management of Iron Steamships—The SS *Xantho* (1872)." *Proceedings* (Fifth National Conference on Engineering Heritage, Perth) (1990): 75–80.

—. "In-Situ Corrosion Studies on Iron and Composite Wrecks in South Australian Waters: Implications for Site Managers and Cultural Toursim." *Bulletin of the Australian Institute for Maritime Archaeology* Vol. 22, 1998. [ECU-784]

MacLeod, Ian Donald and Neil A. North. "350 Years of Marine Corrosion in Western Australia." *Corrosion Australasia* 5:3 (1980): 11–15.

"Making a Meal of Iron." *Science News* 132 (1987): 104. [ECU-106]

Mardikian, Paul and René David. "Conservation of a French Pistol from the Wreck of *Le Cygne* (1808)." *Studies in Conservation* Vol. 41 No. 3, 1996, pp. 161–169. [ECU-673]

Morris, R. "Ferrous Clib Concretion on Small Gems." *International Journal of Nautical Archaeology* 13:1 (1984): 65–70.

Moyer, Curt. "Archaeological Conservation Forum." *Newsletter* (Society for Historical Archaeology)(1990): 19–21. [ECU-305]

"Navy Board's Report to the Admiralty on the First Coppering Experiment—31 August, 1763." *American Neptune* I (1941): 304–306. [ECU-116]

North, Neil A. "Formation of Coral Concretions on Marin Iron." *International Journal of Nautical Archaeology* 5:3 (1976): 253–256. [ECU-61]

—. "Electrolysis of Marine Iron." *Proceedings* (First Southern Hemisphere Conference on Maritime Archaeology, Melbourne) (1977): 145–147.

—. "Corrosion Products on Marine Iron." *Studies in Conservation* 27:2 (1982): 75–83. [ECU-222]

—. "Electrolysis of Marine Iron." *Proceedings* (First Southern Hemisphere Conference on Maritime Archaeology, Melbourne) (1977): 145–147.

—. "The Role of Galvanic Couples in the Corrosion of Shipwreck Metals." *International Journal of Nautical Archaeology* 13 (1984): 133–136. [ECU-62, 595]

North, Neil A. and M. Owens. "Design and Operation of a Furnace for H 2 Reduction of Marine Iron." *International Journal of Nautical Archaeology* 10:2 (1981): 95–100. [ECU-63]

North, Neil A., M. Owens, and Colin Pearson. "Thermal Stability of Cast and Wrought Marine Iron." *Studies in Conservation* 21 (1976): 192–197. [ECU-387]

North, Neil A. and Colin Pearson. "Alkaline Sulfite Reduction Treatment of Marine Iron," *Proceedings* (ICOM Comnittee for Conservation, 4th Meeting) (1975).

—. "Investigations into Methods for Conserving Iron Relics Recovered from the Sea." *Conservation in Archaeology and the Applied Arts* (IIC, London) (1975): 173–181.

—. "Thermal Decomposition of FeOCl and Marine Cast Iron Corrosion Products." *Studies in Conservation* 21 (1977): 146–157. [ECU-176]

—. "Washing Methods for Chloride Removal from Marine Iron Artifacts." *Studies in Conservation* 23 (1978): 174–186. [ECU-179]

—. "Recent Advances in the Stabilization of Marine Iron." *Conservation of Iron Objects Found in Salty Environments*, R. Organ, E. Nosek, and J. Lehmann, eds. (Historical Monuments Documentation Center, Warsaw, Poland) (1978): 26–38.

—. "Methods for Treating Marine Iron." *Proceedings* (ICOM Committee for Conservation, 5th Triennial Meeting, Zagreb) (1978): 1–10.

—. "Long Term Corrosion of Wrought and Cast Iron in Seawater." *Proceedings* (6th International Congress on Metallic Corrosion in Sydney 1975, Australian Corrosion Association 6) (1981): 269–283.

Notoya, T. and G.W. Poling. "Protection of Copper by Pretreatment with Benzotriazole." *Corrosion* 35:5 (1979): 193–200. [ECU-151]

O'Donnell, E.B. and Maureen M. Julian. "Conservation of Two Brass Gudgeons from the 1781 Wreck of HMS *Culloden*." *Proceedings* (16th Conference on Underwater Archaeology) (1985): 79–80. [ECU-68]

Oddy, W.A. "Toxicity of Benzotriazole." *Studies in Conservation* 19:3 (1974): 188–189. [ECU-340]

—. "A Review of Procedures for the Conservation of Cast and Wrought Iron Found on the Sea Bed." *International Journal of Nautical Archaeology* 4 (1975): 367–370. [ECU-64]

Oddy, W.A. and M.J. Hughes. "The Stabilization of 'Active' Bronze and Iron Antiquities by the Use of Sodium Sesquicarbonate." *Studies in Conservation* 15 (1970): 183–189. [ECU-390]

Organ, Robert M. "The Consolidation of Fragile Metallic Objects." *Recent Advances in Conservation* G. Thanson, ed.(IIC, London) (1963): 128–134.

Organ, Robert M. "Conservation of Iron Objects." *Historical Archaeology* 1 (1967): 52–54. [ECU-383]

Parr, J. Gordon. "The Sinking of the *Ma Robert*: An Excursion into mid-19th Century Steelmaking." *Technology and Culture* 13 (1972): 209–225. [ECU-246]

Pascoe, M.W. "Final Discussion." *Conservation of Iron* (1982): 68–69. [ECU-0283]

—. "Organic Coatings for Iron: A Review of Methods," *Conservation of Iron* (1982): 56–57. [ECU-285]

Patscheider, J. and S. Veprek. "Application of Low-pressure Hydrogen Plasma to the Conservation of Ancient Iron." *Studies in Conservation* 31:1 (1986): 29–37. [ECU-190]

Pearson, Colin. "Cannon Survive 200 Years Under the Sea." *Foundry Trade Journal* 132:2882 (1971): 307–310. [ECU-143]

—. "The Conservation of Metals." *The Conservation of Marine Archaeological Objects*. Butterworths, London 1987. ECU-567]

—. "The Preservation of Iron Cannon after 200 Years Under the Sea." *Studies in Conservation* 17 (1972): 91–110. [ECU-73]

—. "Restoration of Cannon and Other Relics from HMS *Endeavour*." (Department of Supply, Australian Defence Scientific Service Defense Standards Lab, Report No. 508) [n.d.] [ECU-72]

Pearson, Colin and Neil A. North. "Methods for Treating Marine Iron." *Proceedings* (International Committee for Conservation, IIC, 5th Triennial Meeting, Zagreb) (1978).

Pelikan, J.B. "Conservation of Iron with Tannin." *Studies in Conservation* 11 (1966): 109–114. [ECU-0389]

Pourbaix, Marcel. "Significance of Protection Potential in Pitting Intergranular Corrosion and Stress-Corrosion Cracking." *Journal of Less Common Metals* 28 (1972): 51–69. [ECU-142]

Rees-Jones, S.G. "Some Aspects of Conservation of Iron Objects from the Sea." *Studies in Conservation* 17:1 (1972). [ECU-0334]

"Restoring Old Ironsides." *National Geographic* Vol. 191 No. 6, 1997. [ECU-785]

Rinuy, A. and F. Schweizer. "Application of the Alkaline Sulphite Treatment to Archaeological Iron: A Comparitive Study of Different Desalination Methods." *Conservation of Iron* (1982): 44–49. [ECU-287]

Robinson, Wendy S. "Observations on the Preservation of Archaeological Wrecks and Metals in Marine Environments." *International Journal of Nautical Archaeology* 10:1 (1981): 3–14. [ECU-418]

—. "The Corrosion and Preservation of Ancient Metals from Marine Sites." *Nautical Archaeology*. Vol. 11 No. 3, 1982: 221–231. [ECU-574]

Rodgers, Bradley A. "Conservation of the Chesapeake Bay Mystery Gun," *Underwater Archaeology Proceedings*, Society for Historical Archaeology Conference (1989): 134–137. [ECU-80]

—. "The Case for Biologically Induced Corrosion at the Yorktown Shipwreck Archaeological Site," *International Journal of Nautical Archaeology* 18:4 (1989): 335–340. [ECU-0079]

—. *The East Carolina University Conservator's Cookbook: A Methodological Approach to the Conservation of Water Soaked Artifacts*. Herbert, R. Paschal Memorial Fund Publication, East Carolina University, Program in Maritime History and Underwater Research, 1992. [ECU-402]

Sanders, John W. "Swivel Guns of Southeast Asia." *Gun Digest*. Ken Warner, ed. Northfield, IL: DBI Books, Inc., 1982, pp. 68–75. [ECU-533]

Schrier, L.L., ed. "Tannins to Control Corrosion." *New Scientist* (1961): 403.

—. *Corrosion*, Vol. 1, *Corrosion of Metals and Alloys*. John Wiley: New York, 1963.

Schwarzer, J. and E.C. Deal. "A Sword Hilt from the Serce Limon Shipwreck." *MASCA* 4:2 (1986): 2–50.

Scott, David A. and N.J. Seely. "The Washing of Fragile Iron Artifacts." *Studies in Conservation* 32 (1987): 73–76. [ECU-384]

Sease, Catherine. "Benzotriazole: A Review for Conservators." *Studies in Conservation* 23 (1978): 76–85. [ECU-192]

Selwyn, L.S., P.J. Sirois, and V. Argyropoulos. "The Corrosion of Excavated Archaeological Iron With Details on Weeping and Akaganéite." *Studies in Conservation* Vol. 44 No. 4, 1999, pp. 217–232. [ECU-674]

Selwyn, L.S., W.R. McKinnon, and V. Argyropoulos. "Models for Chloride Ion Diffusion in Archaeological Iron." *Studies in Conservation* Vol. 46 No. 2, 2001, pp. 109–120. [ECU-675]

Semczak, Carl M. "A Comparison of Chloride Tests." *Studies in Conservation* 22 (1977): 40–41. [ECU-126]

Skinner, Theo. "The Conservation of Some Cast Iron Amunition by the Alkaline Sulphite Method." (The Laboratories of the National Museum of Antiquities of Scotland II) [n.d.]: 1–19.

Smith, C.A. "Early Corrosion Investigations." *Corrosion Protection and Control* 24:1 (1973): 6–13. [ECU-147]

Smith, C. Wayne. "Iconography of the Archangel Michael on Pail Weights from Excavations at Port Royal Jamaica." *International Journal of Nautical Archaeology* Vol. 28 No. 4, 1999. [ECU-786]

South, Stanley A. "A Method of Cleaning Iron Artifacts." *Newsletter of the Southeastern Archaeological Conference* 9:1 (1962): 17–18. [ECU-419]

South, Stanley A. "Notes on Treatment Methods for the Preservation of Iron and Wooden Objects." (Unpublished manuscript, North Carolina Department of Archives and History, Brunswick Town) [n.d.]. [ECU-420]

Southwell, C.R., J.D. Bultman, and A.L. Alexander. "Corrosion of Metals in Tropical Environments–Final Report of 16-Year Exposures." *Material Performance* 15:7 (1976): 9–25. [ECU-152]

Staicopolus, D.N. "The Role of Cementite in the Acidic Corrosion of Iron." *Journal of the Electrochemical Society* 110:11 (1963): 1121–1124. [ECU-273]

Stone, George C. *A Glossary of the Construction, Decoration, and Use of Arms and Armor in All Countries and in All Times*. Portland, ME: The Southworth Press, 1934. [ECU-534]

Tower, Howard B., Jr. "Conserving Metal Artifacts: A Positive Solution in a Negative Situation." *Skin Diver* (July 1989): 19, 180–184. [ECU-425]

Turgoose, S. "The Nature of Surviving Iron Objects." *Conservation of Iron* (1982): 1–7. [ECU-278]

Tylecote, R.F. *A History of Metallurgy*, London: The Metals Society, (1976).

Tylecote, R.F. and J.W.B. Black. "The Effects of Hydrogen Reduction on the Properties of Ferrous Materials." *Studies in Conservation* 25 (1980): 87–96. [ECU-175]

Tyler, J.C. "The Recovery of the *Endeavour's* Cannons." *Australian Natural History* 16:8 (1969): 281–288.

U.S. Government Printing Office. "Corrosion and Metal Artifacts: A Dialoque Between Conservators and Archaeologists and Corrosion Scientists." (NBS Special Publication 479, U.S. Government Printing office, Washington D.C.) [n.d.].

Various. "Archaeological Conservation Forums: 'The Curse of Iron—current research,' In Massachusetts—Can the curse be lifted, Urban Archaeology." SHA Newsletter Vol. 19 No. 4, December 1986. [ECU-569]

Videla, H.A. and M.F.L. de Mele. "Microfouling of Several Metal Surfaces in Polluted Sea Water and Its Relation with Corrosion." *Corrosion 87* (Paper No. 365, NACE, San Francisco) (1987). [ECU-0490]

Vincent, D. "Successful Conservation of *De Surville* Anchors." (Art Galleries and Museums Associations of New Zealand) 8:1 (1977): 4–5.

von Wolzogen Kuhr, C.A.H. and L.S. Vander Vlugt. "The Graphitization of Cast Iron as an Electrobiochemical Process in Anaerobic Soils." *Water* 18:147 (1934).

"The Role of Corrosion Inhibitors in the Conservation of Iron." *Conservation of Iron* (1982): 58–67. [ECU-284]

Walker, Robert. "The Instability of Iron Sulfides on Recently Excavated Artifacts." *Studies in Conservation* Vol. 46 No. 2, 2001, pp. 141–152. [ECU-676]

Washington Cannon Paperwork and Research. [ECU-604]

Watkinson, David. "An Assesment of Lithium Hydroxide and Sodium Hydroxide Treatments for Archaeological Ironwork." *Conservation of Iron* (1982): 28–43. [ECU-288]

Weaver, Martin. "Fighting Rust." *The Journal of the Association for Preservation Technology* 9:1 (1987): 16–18. [ECU-99]

White, John R. "Cleaning Heavily Encrusted Iron Artifacts by Cacination and Induced Spalling." *Curator* 19:4 (1976): 311–315. [ECU-150]

Williams, M.E., R.A. King, and J.D.A. Miller. "Sulfate-Reducing Bacteria, Surface Coatings, and Corrosion." *Journal of the Oil and Colour Chemistry Association* 56:8 (1973): 363–368. [ECU-0134]

Wranglen, Gosta. *An Introduction to Corrosion and Protection of Metals*. New York: Chapman and Hall, 1985.

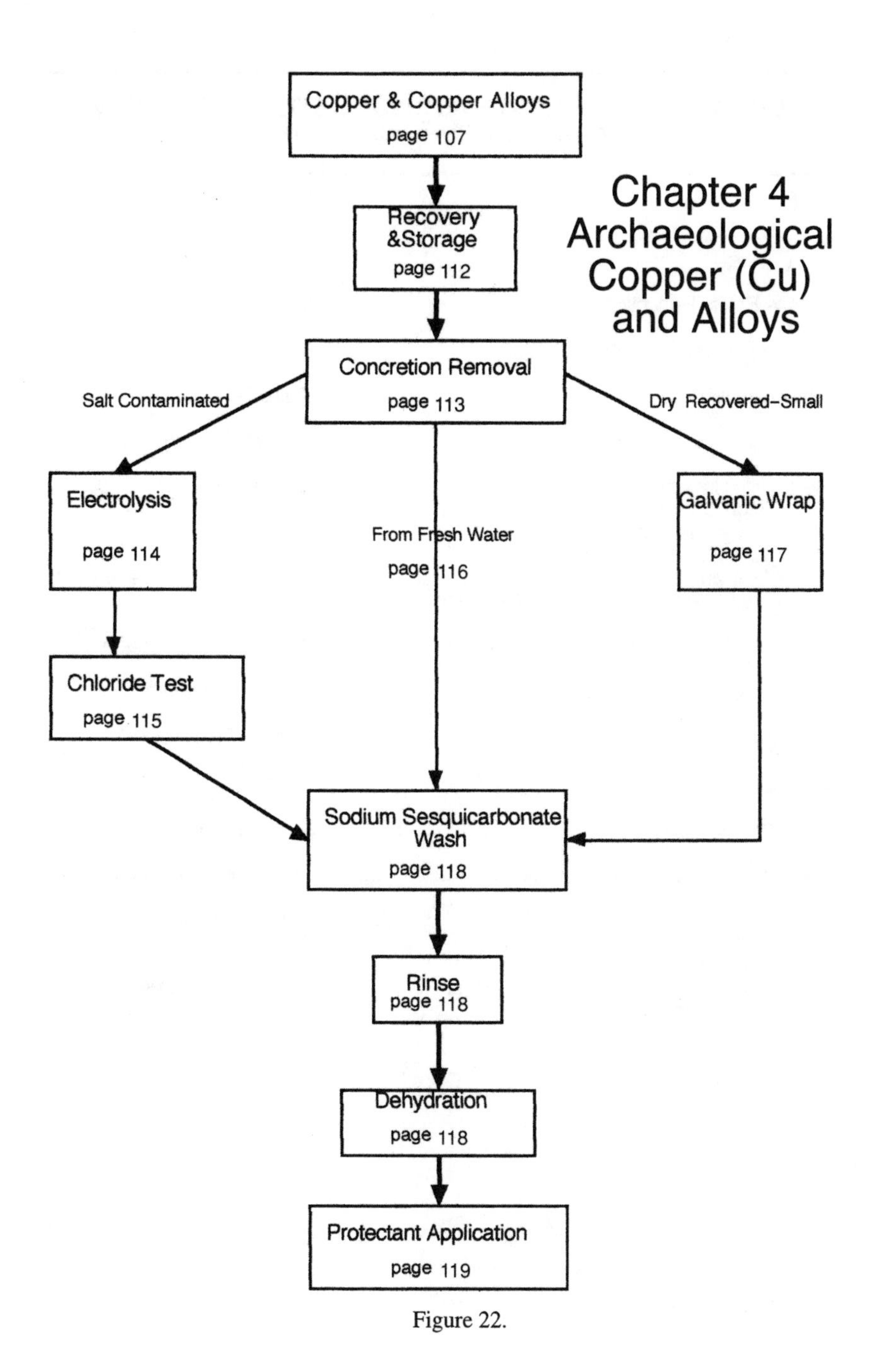

Copper & Copper Alloys
page 107
Recovery
&Storage
page 112
Chapter 4
Archaeological
Copper (Cu)
and Alloys
Concretion Removal
page 113
Salt Contaminated
Dry Recovered-Small
Electrolysis
page 114
From Fresh Water
page 116
Galvanic Wrap
page 117
Chloride Test
page 115
Sodium Sesquicarbonate
Wash
page 118
Rinse
page 118
Dehydration
page 118
Protectant Application
page 119

Figure 22.

Treatments for Archaeological Copper and Alloys

Represented in Conservation Literature

Alkaline Sulfite Reduction..Toxic Factors

Alkaline Sodium Dithionite..Toxic Factors

Amonia (Liquid)..Toxic Factors

EDTA..Surface Treatment

Electrolytic Reduction...Can Work with Light Electrolyte
- High Current Density
- Moderate Current Density
- Low Current Density
- Low amperage/ Light Electrolyte..........................Recommended (page 114)
- Higher Amperage/ Light Electrolyte

Galvanic Wrap (Electrochemical Cleaning).......................Recommended (page 117)

Phosphoric Acid...Surface Treatment

Mechanical Surface Cleaning..Surface Treatment

Orthophosphoric Acid...Surface Treatment

Sequestering Agents.. ....Surface Treatment

Sodium Carbonate Wash...Slow, blocked by corrosion

Stripping Acids..Surface Treatment
- Oxalic
- Thioglycollic
- Citric..Will clean concretion
  - Modalene
  - Biox

Figure 23.

Chapter 4

Archaeological Copper (Cu) and Copper Alloys

COPPER—THEORY

Copper may rival gold as the first metal used by man. Its distinct red color and blue green oxides are unmistakable, no doubt catching the notice of our late stone-age ancestors. Unlike iron, native copper can sometimes be found in its metallic form. Copper has many properties that made it useful. It is malleable, ductal, and resists corrosion, native copper is easily worked and hammered into useful tools and ornaments.

Copper use appears in Asia Minor around 8000 BC but the production of copper from ore does not begin until the middle of the fourth millennium BC. Modern copper extraction and smelting is a complex process specifically engineered for the ore type to be refined. However, historic copper smelting from high-grade ores or the mineralized oxides of copper was relatively easy. Copper has a low melting point and could be reduced in an open fire and later in open-hearth furnaces. Its oxidized ores such as nautokite, paratacamite, and cuprite were easy to identify by their blue and blue/green colors, but the most plentiful ores are the anaerobic sulfide minerals such as chalcocite and covellite that are dark in color.

Copper has many properties that allow it to survive well in an archaeological environment. Unlike iron, copper is fairly stable in metallic form. When it does corrode the corrosion product, or patina, is electrically passive. The electrically passive patina polarizes the anode, greatly slowing the corrosion rate. The ions of this metal are poisonous to most life forms greatly altering how concretions form on copper compared to non-poisonous iron. So copper artifacts can easily survive thousands of years in the soil or sea floor environment.

Native copper and copper ores generally contain other metals such as lead, nickel, zinc, gold, silver, platinum, arsenic, and antimony as well as other minerals such as phosphorous. The trace amounts of these contaminants can sometimes be used to reveal the source of native copper or the areas from which historic ores have been extracted. The trace metals also add to copper's corrosion resistance by greatly slowing the electron flow from the metal.

Pure copper is too soft to make lasting edges or cutting tools. In its pure form it was not much more useful than a knapped bi-face stone tool. Yet around the time copper was first smelted from ore, early metallurgists found that combining it with various amounts of tin greatly improved its hardness. Bronze, the alloy of copper and tin, dominates as the most useful metal for two millennia until iron is introduced.

Bronze contains various amounts of tin depending on the desired properties of the metal. Bronzes that contained only about 10% tin could be worked in a forge to produce fasteners and tools, much as wrought iron was later used. Bronze can contain 25% or more tin plus various lesser percentages of other metals. Historic bronze can also contain iron or lead, in which case the alloys are referred to as iron or leaded bronzes. Bronze could be fashioned into tools, statues, cannon, bells, and swords.

By the early Iron Age metallurgists discovered that they could combine copper with zinc to form a more malleable and easily worked alloy called brass. Brass can be worked in a forge similar to wrought iron, or it can be poured into molds. Brass contains many of the trace elements found in bronze including arsenic, antimony, and phosphorous, all of which tend to inhibit electron flow and corrosion. The alloy can contain up to 45% zinc, making it harder yet more brittle than pure copper. Brass can also contain tin, nickel, lead, and iron in varying amounts. Some of these alloying metals are added to change the properties of the metal, some are contaminants, and others like iron, are added to make the brass cheaper and less expensive to produce. Though copper can be found in metallic form and its early use for tools and ornaments is easily understandable, it is rarer and more expensive than its alloying metals.

Brass, can be divided into two major categories, High and Low brass. High brass is actually a cheaper metal containing over 32% zinc plus other metals. It is bright yellow in color and can be mistaken on sight alone for gold. High Brass is also know variously as Yellow Brass, Cartridge Brass, or Muntz Metal often depending on its use. High brass can be used for fasteners and sheathing for wooden ships, for ornaments and jewelry, and as the name implies, cartridges by mid-19th century.

Low Brass is 20%–30% zinc, and is less reactive or more corrosion resistant the high brass. Low Brass is also variously known as Admiralty Brass or Red Brass. This brass is more malleable than high brass and is darker in color, very similar to pure copper. Low brass's corrosion resistance made for ideal ship's fittings, pipes, engine fittings and gauges.

CORROSION IN COPPER AND ALLOYS—THEORY

Like all metals, metallic copper will corrode in order to seek its' most stable electrochemical form. Usually this means that a metal will return to the perdurable oxide or sulfide ores that it was originally smelted from, turning full circle electrochemically. In this light, copper will begin to corrode in a similar way to iron by acquiring a positive charge and becoming an anode with a galvanic potential.

Anode (+)

$$Cu^0 \rightarrow Cu^{+2} + 2e^-$$

$$4Cu^0 + 2H_2O + O_2 + 4e^- \rightarrow 4CuO + 4H^+$$

As with iron, this formula means that metallic copper will give off electrons. The electron discharge allows the copper ions to migrate from the surface of the metal and combine with water and oxygen to form the copper oxides (Cuprite and or Tenorite) and acid. In the presence of chloride anions the reaction and outcomes are nearly the same with the formation of cuprous chloride is CuCl, cupric chloride (nautokite) $CuCl_2$ or cupric hydroxy chloride (paratacamite) $Cu_2(OH)_3Cl$.

At the cathode the electrons given off by the copper reduce water and oxygen to form alkaline hydroxyl ions.

Cathode (−)

$$O_2 + 2H_2O + 4e^- \rightarrow 4OH^-$$

Yet there are complications in copper corrosion not found in iron corrosion. Greatest among these is the fact that the corrosion products do not conduct electricity. This automatically polarizes the reaction by placing a barrier between the positive and negative pole of the corrosion reaction. If the exchange of electrons stops, the corrosion will also cease. Pure copper then will only corrode so long as the surface is cleaned of this patina by physical wear or weathering.

Yet so long as corrosion is an electrical transfer, differential aeration corrosion will still be a problem with copper, as it is with iron. For the anode (area of disassociation of the metal) may not be the area where the electrons are reducing water and oxygen. So corrosion product build up at the anode may not prevent electrical exchange at the cathode.

Copper fasteners will show this corrosion in the form of a narrowing of the artifact or necked area. The necking will demonstrate to the archaeological observer where the metal was once covered, even if that covering (wood for instance) has long since disappeared.

Alloys of copper pose a greater corrosion risk than does pure copper. It is physically impossible to completely mix molten metals. Impurities and other contaminants also contribute to a less than perfect alloy. This will create areas on the surface of an artifact with a higher corrosion potential than the surrounding metallic surface. The metal ions in this area will begin to give off electrons and begin to migrate as free charged ions before combining with anions such as chlorides or oxygen to form corrosion products. The charge given off by the corroding anodic area is absorbed by adjacent metal. This metal will act as the cathode for the reaction and water and oxygen will be reduced to hydroxyl ions at the surface. The result is pitting corrosion or the dissolution of metal from one defined place on the surface of an object.

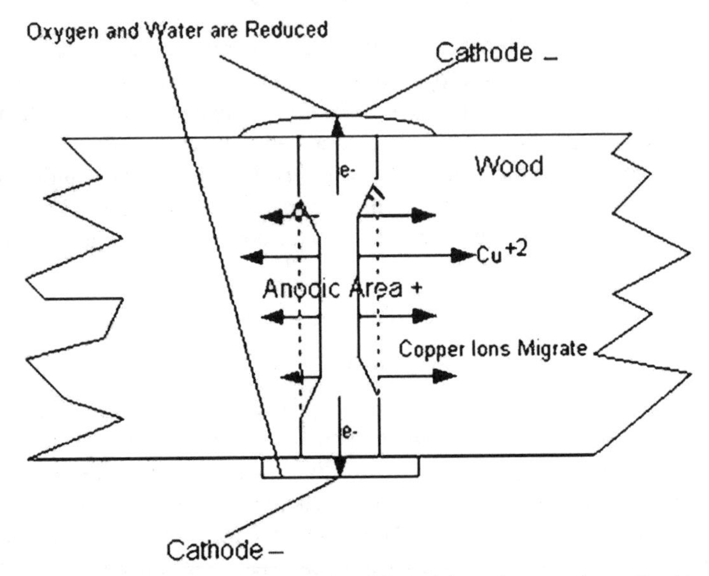

Figure 24. Copper alloy rivet undergoing differential aeration corrosion or "necking".

Pitting is a concern in both the preservation and storage of copper alloy artifacts. Its onset is often seen as mysterious and its effects on the surface of the artifact can be devastating if unchecked.

Yet pitting can be a minor concern in alloys. Particularly alloys with as great a Corrosion Potential (E Corr) difference as there is between copper and tin (with bronze) and copper and zinc (with brass). Normally these metals lying in electrical contact would initiate a galvanic coupling (see galvanic corrosion page 76) in which the less noble metals would donate electrons and corrode while reducing the more noble metals, which will be protected. And of course galvanic coupling is a concern on any archaeological site that might contain dissimilar metals. Almost invariably when iron and copper are in contact the iron will be corroded and the copper will look, if not pristine, then very good comparatively.

This same galvanic process does occur inside bronze and brass. The tin in the case of bronze or the zinc in the case of brass will selectively corrode out of the alloy. Dezincification, in the case of brass, or destanification, in the case of bronze leaves a matrix of crystalline copper where the alloy used to be. Dezincification and destanification are problems that are mainly encountered in sea water but could also be a problem in earth bound artifacts. Galvanic dissolution will reduce the

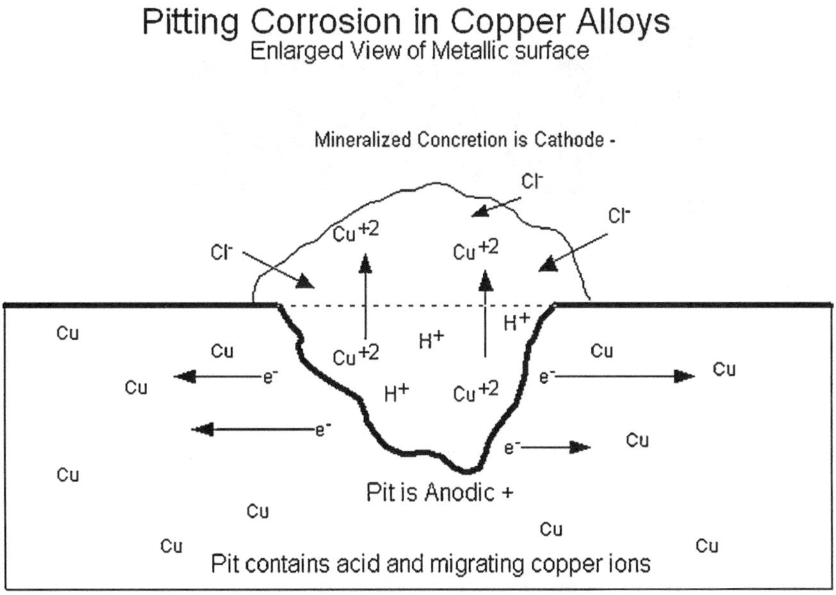

Figure 25. Copper alloy undergoing pitting corrosion. Pitting is problematic even in storage.

Figure 26. Dezincified brass bolt. The zinc has preferentially corroded weakening the bolt while leaving a crystalline copper matrix.

physical strength of an artifact. A dezincified brass bolt, for instance, will have no more strength than a piece of chalk. Great care should be taken in handling copper alloys that are suspected of selective corrosion attack.

CONCRETION FORMATION—THEORY

Dry or earth recovered copper artifacts will rarely have much concretion only a surface layer of corrosion product. Dark brown or black corrosion products will indicate anaerobic corrosion with dominant sulfide staining. Blue and green

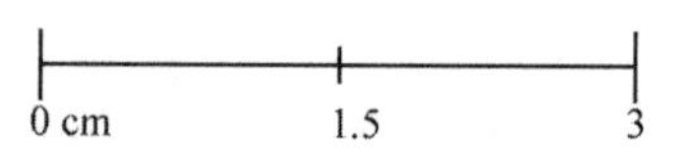

Figure 27. Copper alloy concretion begins with the deposit of calcium carbonate on an alkaline surface just above the corrosion product. Afterward, shellfish can attach themselves to the artifact (Note C shaped shell attachment).

corrosion products will indicate oxides produced in shallow burials. Unlike iron, there are methods for reducing the corrosion products on the surface of copper and alloyed artifacts back to a metallic form, so the corrosion should not be removed under any circumstance during archaeological recovery.

In the sea, concretion can form on copper artifacts, though it will never be as thick as it is on iron (see iron concretion). Copper ions are toxic to the fouling organisms that would normally colonize a metallic artifact on the sea floor making copper the natural choice in the late 18th century for shipbuilders to sheath their wooden ships. But the initial copper corrosion can be enough to acidify the layer of water near the artifact's surface. The acidification of this layer will supersaturate the water there in calcium carbonate. As this water moves away from the artifact the calcium carbonate may precipitate out of the water forming a white chalky shell on the artifact. In time fouling organisms may be able to colonize the calcium carbonate surface of the artifact, but never to any depth. There is no indication that sulfate reducers or methanogens can colonize the interior of these concretions. Green, blue/green migrating copper ions will seep to the surface of the concretion and inhibit most continued coral growth. As on land, dark brown and black sulfide stains will indicate that an artifact was interred in the anoxic regions below the sea bottom.

RECOVERY AND STORAGE OF COPPER AND ALLOYS—METHODOLOGY

Dry site recovered copper and copper alloy artifacts, like all metallic artifacts on recovery, should be placed into an aqueous basic solution of 2% to 5% sodium carbonate or sodium bicarbonate (baking soda). The solution is easily mixed with 2 to 5 grams of sodium carbonate or bicarbonate added to 98 to 95 ml of distilled or deionized water (see storage of iron). Damp concretions that are allowed to air dry do seem to become tougher than if they are immediately submerged in an alkaline solution. The artifacts should also be lightly wrapped in aluminum foil that lets the storage solution penetrate to the artifact. This will protect the artifacts from oxidation while they are in storage. The combination of aluminum foil and sodium carbonate will begin the reduction process that will continue in the laboratory with electrolysis or galvanic wrapping.

Storage of dry recovered artifacts should be in the same basic solution where they were placed on recovery. Not only will the alkaline solution and aluminum foil protect the artifacts, electrochemically, the water itself and the buoyancy it affords will mitigate some of the effects of gravity on the most friable artifacts.

Most copper and copper alloy artifacts recovered from the sea are in a good state of preservation, covered in a blue/ green patina. They will, however, be chloride contaminated and need thorough electrolytic reduction. Copper and alloys recovered from fresh water are in better condition, sometimes as shiny as the day they were cast or wrought.

Storage is the same for sea and fresh water recovered artifact as it is for dry site recovered artifacts. Immediately on recovery they should be wrapped in aluminum foil and placed in a 2% to 5% sodium carbonate or sodium bicarbonate solution. This will begin the reduction process and mitigate any damage done to the artifact on its removal from the site.

Copper alloy artifacts recovered from anaerobic environments in the benthic (sea floor) environment are invariably in the worst condition of those mentioned above. Attacked by sulfate reducers, these artifacts will be covered with a black or dark brown sulfide layer of chalcolite or covellite. These surface coating are apparently not as electrically passive as the oxide coatings, making it even more of a priority that the artifact be stored in an alkaline solution and wrapped in aluminum foil for galvanic reduction and protection.

CONCRETION REMOVAL—METHODOLOGY

Concretion in this section should be minimal. It is defined as dirt, carbonate, and calcium carbonate adhering to the artifact. The dirt can be scrubbed cleaned under a stream of fresh water with soft nylon bristle brushes and a paste of sodium bicarbonate in water.

The calcium carbonate will disappear during electrolysis or the galvanic wrap phase, along with the corrosion product. If it remains stubborn despite the reduction phase it is easily removed in a citric acid soak. To accomplish this an artifact is placed in a container of 10% weight to volume citric acid. This is a mixture of 10 grams of citric acid crystals in 90 ml of water. This acid is safe at this concentration and in fact will exhibit biotic growth over time. The growth does not interfere with the performance of the acid. Citric soaks can last from an hour to several days.

It is not, however, advisable to immediately acid soak an artifact before treatment as this will remove some of the corrosion product. Electrolysis and galvanic wrapping will actually reduce this corrosion back to elemental metal. This can bring out previously unseen fabrication marks, delicate etchings, not previously seen on machined or stamped plates.

ELECTROLYTIC REDUCTION FOR SALT CONTAMINATED ARTIFACTS—METHODOLOGY

Electrolytic reduction of copper alloys follows the same basic procedure outlined for iron electrolysis. Again, as with iron, the main concept is to reduce the metal corrosion products to more stable forms, and thereby increase the porosity of the artifact in order that the chlorides rinse faster. Unlike iron, however, there is evidence that some of the corrosion products in copper and copper alloys will reduce back to elemental copper, so it is important that the corrosion products not be removed before reduction. Should these corrosion products, or the calcium carbonate concretion remain stubbornly attached to the artifact, the concretion removal segment should be revisited.

Most salt contaminated brass and bronze artifacts respond well to low current density electrolytic reduction. Amperages should be limited to less than two amps except for large objects that can take more current. A *light* electrolytic solution of sodium carbonate is safe for all metals and copper alloys are no exception.

Set up for copper and copper alloyed artifacts is the same as it is for iron artifacts with one exception, the sacrificial anodes should be copper. The object is placed in a container submerged in .25% to .5% sodium carbonate or sodium bicarbonate electrolyte. The artifact is grounded via an alligator clip to the negative terminal of the DC power source and surrounded by sacrificial copper anodes connected to the positive terminal of the power source. At low current density and low amperage a certain amount of metal from the anodes will plate out on the artifact and corrosion residue in the surface inter-granular boundary will reduce to metallic copper. This can be a great help in defining wear patterns, fabrication marks, or illegible designs and inscriptions in the metal's surface for all of these will be brought to crystal clarity.

Since plating will occur on the artifact it is suggested that the anodes for copper and its alloys should also be copper, brass, or bronze. This is not a mandatory step but will save some work in the long run. If mild steel anodes are used, magnetite will plate out on the copper alloy artifact. Since magnetite is black it will need to be scrubbed off of the surface of the artifact to bring back the proper coloration. This is easily done with a tooth brush and sodium bicarbonate paste, but takes some time.

The copper alloy sacrificial anodes can be scrap metal or copper piping from a hardware store. It should be realized, however, that the alloying metal will be the first to corrode in the anode. This means the artifact could be plated with zinc, tin, and even magnetite as there is a good deal of iron in modern copper pipes. All of these plated coatings can be scrubbed off the artifact and it is not a great problem. The best modern copper to use for anodes is thick wire. Wire grade copper is nearly pure and will plate on the artifact as pure copper, making the alloy artifact look as though it is made of pure red copper. Though this plating is unrealistic in color an aesthetically pleasing and natural patina is returned through finishing soaks in sodium sequicarbonate. Copper wire anodes are flexible, easy to use, and can be arranged to lye close to the artifact. Their only drawback is that they will need to be changed often, as they will corrode quickly.

The anodes should be placed so they rise above the surface of the electrolyte, at least in one area so they can be connected to the positive terminal of the DC power source without submerging the connecting alligator clips. If the clips are submerged they will corrode in a matter of minutes and the electrolysis will cease.

Electrolytic reduction should be initiated with 6 volts but can be increased to 12 volts to produce the desired amperage. If all is working properly the conservator will notice a blue pigmentation near the anodes in the electrolytic solution almost immediately on turning on the direct current power. As would be expected in a mildly basic electrolyte, the anodes will corrode inundating the solution with a light blue color. The blue copper ions and oxides will diffuse throughout the electrolytic solution and obscure the view of the artifact. Hydrogen and oxygen evolution at the artifact and anodes will foam the surface of the water. The blue pigmentation will settle in three days and the artifact will again be in view.

After the electrolytic reduction is complete, care should be taken in the disposal of the electrolyte. Unlike iron, the copper solution has some mild toxicity and should not simply be sent down the drain. Instead, the electrolyte should be allowed to evaporate and the oxides removed and disposed of properly.

CHLORIDE TEST—METHODOLOGY

Chloride testing for copper alloys is identical to chloride testing for iron artifacts. The testing is not done for every artifact, only those large enough for a chloride measurement to be meaningful. A small sample of electrolyte is placed in a 50 ml beaker and a titrator strip inserted. After the solution has wicked to the top, the white line on the strip is read and the chloride concentration calculated from the chart included with the titration test. As in larger iron artifacts a weekly reading of chlorides should be plotted against time. This graph acts as a control over solution change and as an indication when the overall chloride concentration

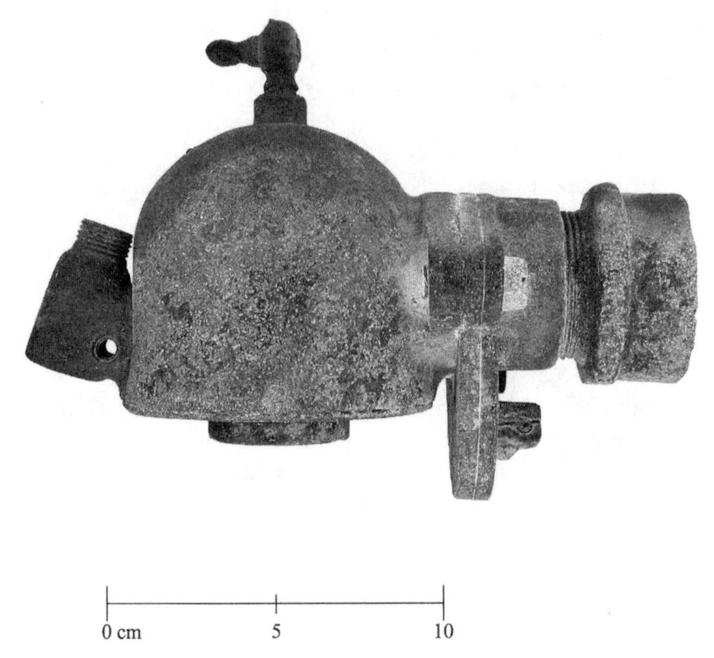

0 cm 5 10

Figure 28. Turn of the 20th century fishing boat brass carburetor with generalized corrosion and salt intrusion between the flanges. Photograph by Chris Valvano.

has been lowered within acceptable limits, a leveling near 50 ppm with a solution change at 1000 ppm. Copper alloys, however, do not absorb nearly the amount of chlorides that iron artifacts will. For this reason the measurement of chloride ion concentration in the electrolyte will not show the dramatic changes that is does in iron, particularly cast iron.

Smaller copper and copper alloy artifacts may not release enough chlorides for any meaningful measurement to take place. Pins, tacks, brads, and small items will take one to two weeks. Fastener sized artifacts like large nails will take three to four weeks. Larger artifacts like drift pins will take longer, up to six weeks. As with iron, the only real test of whether an artifact has received enough treatment is whether it begins to corrode in storage. Artifacts made from different attached sections may have to be taken apart so the rinse process can clean the chlorides from all of the cracks and crevices.

FRESH WATER ARTIFACTS—METHODOLOGY

Fresh water recovered artifact can be those defined as recovered from fresh water lakes, rivers, or ponds. These artifacts are oftentimes in great condition and need no special treatment. They should be recovered and stored, as are all metal artifacts recovered from archaeological sites, in a basic solution wrapped in aluminum foil. This is likely be all the treatment these artifacts will need until their sesquicarbonate wash.

GALVANIC WRAP FOR DRY RECOVERED SMALL ARTIFACTS—METHODOLOGY

Electrochemical cleaning with a galvanic wrap (see iron) begins for small artifacts after they have been recovered, wrapped in aluminum foil, and placed in sodium carbonate or sodium bicarbonate storage solution. Unlike simple storage, however, this segment of galvanic wrapping involves a change of electrolyte, and an active interaction with the conservator in concretion removal.

As with storage, the galvanic wrap begins with the placing of a small dry recovered artifact into an aluminum foil pouch. The pouch should be two foil layers thick. It has been determined through experience that, although the galvanic wrap in sodium carbonate will work to conserve and reduce the copper or copper alloy artifacts, replacing the sodium carbonate with citric acid gives faster, clearer, results. Should neither of these electrolytes be available, vinegar (acetic acid) will also work.

The electrolyte should be placed into the pouch with the artifact so it can carry the electrons given off by the aluminum to the artifact. Since the aluminum foil is acting as an anode it will corrode and leak, so it should be placed in a suitable container filled with the same electrolyte that is in the pouch. The foil should be compacted around the artifact to be in as close contact as possible, but care should be taken not to crush the artifact.

After several days the foil pouch should be opened and the artifact observed. Most of the corrosion product should now be changed to copper or a brown copper oxide. The surface should be clean and clear. Any carbonate and attached adherent sand and dirt should be gone. If it is not, continue the wrap. There is no definite time period allocated to finishing the galvanic wrap. Generally the effects are

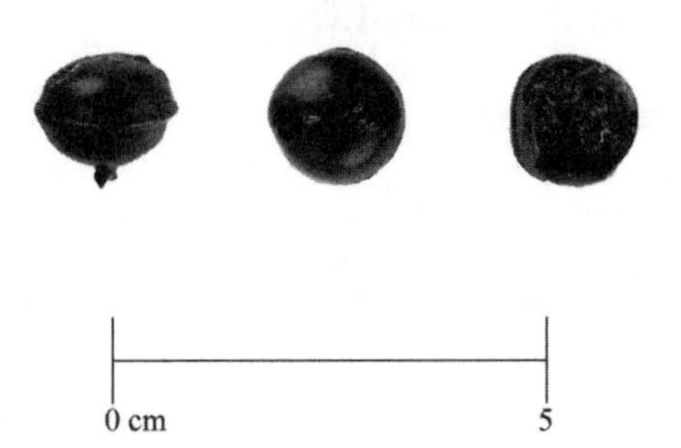

Figure 29. Early 18th century brass bells with iron clappers from Fort Neoheroka. The clapper in the bell on the right has corroded completely in its galvanic coupling with the brass. Photograph by Chris Valvano.

dramatically improved in only a few days. However, the treatment can be kept up indefinitely as there have never been any observable bad effects.

SODIUM SESQUICARBONATE WASH—METHODOLOGY

By this point the artifact will be clean, free of debris, and corrosion products. It may also be shiny and new looking, or copper colored and unnatural, depending on the alloy. It is up to the conservator to decide what the final appearance of the artifact will be. If the artifact is to remain shiny its should be sent to the rinse phase. However during the useful life of most cuprous artifacts, they carried a brown copper oxide patina, like an old penny. This patina can be restored during the sodium sesquicarbonate wash.

This can be a vigorous wash for robust artifacts, or a delicate soak for those that may not be able to withstand the energy of an active wash. Sesquicarbonate 4%, is a mixture of 20 grams of sodium carbonate and 20 grams of sodium bicarbonate in 960 ml distilled water. The cleaned artifact should be completely submerged in this solution for several days. Boiling the solution for periods of two hours at a time will greatly speed the wash and free any harmful anions still residing in the artifact. A soak of two to three weeks with a lengthy boil each week will restore the copper oxide brown color to the treated artifacts. Delicate objects get the same treatment without the boiling.

RINSE—METHODOLOGY

This is the final rinse before dehydration. Artifacts at this point should be handled with rubber gloves or utensils to prevent salt contamination from fingertips. In this stage the artifacts are placed in distilled water and left to soak for several hours to remove the sesquicarbonate.

DEHYDRATION—METHODOLOGY

Oven dehydration, at times, tends to darken and discolor copper and its alloys, often leaving an iridescence on the surface of the artifact. Solvent drying, therefore, is the best approach to use with cuprous alloys.

The same approach is used here as solvent drying for iron. The artifact is transferred to three successive baths of acetone or alcohol. Each bath should last at least one hour but overnight soaks are not a problem. It should be remembered here that solvents are extremely flammable. They should be covered at all times and never located near an ignition source. Solvents can be reused, with the final soak reserved for clean unused solvent.

PROTECTANT APPLICATION—METHODOLOGY

Copper aloys can be micocrystalline wax coated, similar to iron artifacts. The wax is melted and the artifacts dipped in for about two hours past the time it stops effervescing. The burners should be turned to low and the wax is cooled somewhat before the artifact is removed to drip and cool. Excess wax is smoothed or cut off with a knife.

Copper alloy artifacts can also be coated with Incralac, a shellac-like substance containing benzotriazole (BTA). It should be noted that BTA bonds chemically with the alloy and is non-reversible. According to the MSDS (Material Safety Data Sheets) BTA is a carcinogen and its use should be limited to ventilated areas, or outdoors. Conservators using Incralac should protect themselves with gloves and a respirator.

CONCLUSION

In many ways the durable nature of copper and its alloys lends itself to trouble-free conservation. Artifacts of copper and its alloys brass and bronze, are durable and long lasting. They will survive well in the archaeological record and if their corrosion product is reduced and plated once again as pure metal on the artifact they may well envisage the same detail they had when they were in active use. The artifacts detailed micro-excavation should easily provide details of manufacturing and wear that will not be as well preserved on other metals. Yet these details will only be revealed if the artifact is protected and conserved from the moment it is archaeologically located, until after it goes into final storage.

CHAPTER 4: ARCHAEOLOGICAL COPPER (Cu) AND COPPER ALLOYS

American Copper and Brass. New York: Bonanza Books, 1968.

Alsop, Joseph. "Warriors from a Watery Grave." National Geographic 163 (1983): 820–827. [ECU-271]

Angelucci, S., P. Fiorentino, J. Kosinkova, and M. Marabelli. "Pitting Corrosion in Copper and Copper Alloys: Comparative Treatment Tests." *Studies in Conservation* 23 (1978): 147–156. [ECU-127]

Anon. "Navy Board's Report to the Admiralty on the First Coppering Experiment—31 August, 1763," *American Neptune* I (1941): 304–306. [ECU-116]

Bankok National Museum. *Bronze Disease and its Treatment*. Bankok: Department of Fine Arts, Bankok Museum, 1975.

Barclay, B. "Basic Care of Coins and Metals." *CCI Notes* (Canadian Conservation Institute) 9/4 (1991). [ECU-355]

Barker, Richard. "Bronze Cannon Founders: Comemnts upon Gilmartin 1974, 1982." *International Journal of Nautical Archaeology* 12 (1978): 23–27. [ECU-242]

Bianchi, G. and P. Longhi. "Copper in Sea Water, Potential pH Diagrams." *Corrosion Science* 13 (1973): 853–864. [ECU-219]

CCI Laboratory Staff. "The Cleaning, Polishing, and Protective Waxing of Brass and Copper." *CCI Notes* (Canadian Conservation Institute) 9/3 (1988). [ECU- 467]

CCI Laboratory Staff. "Basic Care of Coins and Metals." *CCI Notes* (Canadian Conservation Institute) 9/4 (1989). [ECU-466]

CCI Laboratory Staff. "Silver—Care and Tarnish Removal." *CCI Notes* (Canadian Conservation Institute) 9/7 (1993). [ECU-473]

Cohen, A. "Copper Alloys in Marine Enviroments." (1975 Liberty Bell Corrosion Symposium) (1973).

—. "The Corrosion of Copper-Nickel Alloys in Sulfide-Polluted Seawater: The Effect of Sulfide Concentration." *Corrosion Science* Vol. 34 No. 1, pp. 163–177. [ECU-677]

Cotton, J.B. "Control of Surface Reactions on Copper by Means of Organic Reagents." *Proceedings* (Second International Congress on Metallic Corrosion, New York) (1963): 590–596.

Cottrell, L.H. "An Introduction to Metallurgy: Copper and Its Alloys." *Retrieval of Objects from Archaeological Sites*. Archetype Publications, London. [ECU-568]

Drayman-Weisser, Terry. "A Perspective on the History of the Conservation of Archaeological Copper Alloys in the United States." *Journal of the American Institute for Conservation* Vol. 33 No. 2, 1994. [ECU-678]

Dugdale, I., and J.B. Cotton. "An Electrochemical Investigation on the Prevention of Staining of Copper by Benzotriazole." *Corrosion Science* 3 (1963): 69–74. [ECU-187]

Duncan, S., and H. Ganiaris. "Some Sulphide Corrosion Products on Copper Alloys and Lead Alloys from London Waterfront Sites." *Recent Advances in the Conservation and Analysis of Artifacts* J. Black, ed. (Institute of Archaeology Summer Schools Press, London) (1987): 109–118.

Faltermeier, Robert B. "A Corrosion Inhibitor Test for Copper-Based Artifacts." *Studies in Conservation* Vol. 44 No. 2, 1999, pp. 121–128. [ECU-679]

Fox, Georgia L. "A note on the use of alkaline dithionite for treating ancient bronze artifacts." *Studies in Conservation* Vol. 40 No. 2, 1995, pp. 139–142. [ECU-680]

Fuller, Cbarles. *The Art of Coppersmithing*. New York: David Williams, 1894.

Furer, J., M. Lambertin, and I.C. Colson. "Morphological and Kinetic Study of Copper Corrosion when Covered with Digenite Sulphide Layers, in Sulphur, under Covvelite Formation Conditions." *Corrosion Science* 17 (1977): 625–632. [ECU-496]

Ganorkar, M.C., V. Pandet Rae, P. Gayaltri, and T.A. Sreenivasa-Rae. "A Novel Method for the Conservation of Copper-Based Artifacts," *Studies in Conservation* 33 (1988): 97–101. [ECU-385]

Giuliani, L. and G. Bombara. "An Electrochemical Study of Copper Alloys in Chloride Solutions." *British Corrosion Journal* 5 (1970): 179–183. [ECU-139]

Hamilton, Donny L. *Basic Methods of Conserving Underwater Archaeological Material Culture*. Washington, D.C.: U.S. Department of Defense Legacy Resource Management Program, January 1996. [ECU-610]

Huda, Khatibul. "A note on the efficacy of ethylenediaminetetra-acetic acid disodium salt as a stripping agent for corrosion products of copper." *Studies in Conservation* Vol. 47 No. 3, 2002, pp. 211–216. [ECU-681]

Joseph, G., and M.T. Arce. "Contribution to the Study of Brass Dezincification." *Corrosion Science* 7 (1967): 597–605. [ECU-137]

Keith, Donald H and Worth Carlin. "A bronze cannon from *LaBelle*, 1686: its construction, conservation, and display." *International Journal of Nautical Archaeology* Vol. 26 No. 2, 1997, pp. 144–158. [ECU-682]

—. "A Laboratory Study of Corrosion Reaction on Statue Bronze." *Corrosion Science* Vol. 34 No. 7, pp. 1083–1097. [ECU-683]

LaFontaine, Raymond H. "The Use of a Stabilizing Wax to Protect Brass and Bronze Artifacts." *Journal of the IIC-CG* 4:2 (1980).

Little, Arthur D., Inc. "Protective Coating for Copper Metals." (International Copper Research Association, Inc., Report No. 97) (1972).

MacLeod, Donald Ian. "Bronze Disease: An Electrochemical Explanation." *ICOM Bulletin* 1 (1981): 16–26.

—. "Formation of Marine Concretions on Copper and Its Alloys." *International Journal of Nautical Archaeology* 11:4 (1982): 267–275. [ECU-243]

—. "Conservation of Corroded Copper Alloys: A Comparison of New and Traditional Methods for Removing Chloride Ions." *Studies in Conservation* 33 (1987): 25–40. [ECU-191]

—. "Stabilization of Corroded Copper Alloys: A Study of Corrosion and Desalination Mechanisma." *ICOM* (Committee for Conservation, Sydney) (1987).

MacLeod, Ian Donald and R.J. Taylor. "Corrosion of Bronzes on Shipwrecks—A Comparison of Corrosion Rates Deduced from Shipwreck Material and from Electrochemical Methods." *Corrosion* 41:2 (1985): 100–104.

Madsen, H. Brinch. "A Preliminary Note on the Use of Benzotriazole for Stabilizing Bronze Objects." *Studies in Conservation* 12 (1967):163–166.

—. "Further Remarks on the Use of Benzotriazole for Stabilizing Bronze Objects." *Studies in Conservation* 16 (1971): 120–122.

Mcneil, Michael B. and Brenda J. Little. "Corrosion Mechanisms for Copper and Silver Objects in Near-Surface Environments." *Journal of the American Institute for Conservation* Vol. 31 No. 3, 1992. [ECU-684]

Merk, Linda A. "A Study of Reagents used in the Stripping of Bronzes." *Studies in Conservation* 23 (1978): 15–22. [ECU-184]

Mor, E.D. and A.M. Beccaria. "Behavior of Copper in Artificial Seawater Containing Sulphides." *British Corrosion Journal* 10 (1975): 33–38.

—. "Effects of Hydrostatic Pressure on the Corrosion of Copper in Water." *British Corrosion Journal* 13:3 (1978): 142–146. [ECU-140]

—. "Effects of Temperature on the Corrosion of Copper in Seawater at Different Hydrostatic Pressures." *Wersktoffe und Korrosion* 30 (1979): 554–558.

Nord, Anders G., Karin Lindahl, and Kate Tronner. "A note of spionkopite as a corrosion product on a marine copper find." *Studies in Conservation* Vol. 38 No. 2, 1993, pp. 92–98. [ECU-685]

North, N.A. and Colin Pearson. "Investigations Into Methods for Conserving Iron Relics from Marine Sites." *Contributions to the 1975 Stockholm Conference*. IIC London, 1976: 173–182. [ECU-573, 606]

Pearson, Colin, ed. *Conservation of Marine Archaeological Objects*, London: Butterworths Series in Conservation and Museology, 1987. [ECU-6]

Richey, W.D. "The Interaction of Benzotriazole with Copper Compounds." (ICOM Comittee for Conservation, Madrid) (1972).

Scott, David A. "Ancient Bronzes Modern Science." *Chemistry in Britain* Vol. 31 No. 8, 1995, pp. 627–630. [ECU686]

—. "An Examination of the Patina and Corrosion Morphology of Some Roman Bronzes." *Journal of the American Institute for Conservation* Vol. 33 No. 1, 1994. [ECU-686]

Schussler, A. and H.E. Exner. "The Corrosion of Nickel-Aluminum Bronzes in Seawater I—Protective Layer Formation and the Passivation Mechanism." *Corrosion Science* Vol. 34 No. 11, 1993. [ECU-704]

Schussler, A. and H.E. Exner. "The Corrosion of Nickel-Aluminum Bronzes in Seawater II—The Corrosion Mechanism in the Presence of Sulfide Pollution." *Corrosion Science* Vol. 34 No. 11, 1993. [ECU-705]

Sharma, V.C. and B.V. Kharbade. "Sodium Tripolyphosphate—a safe sequestering agent for the treatment of excavated copper objects." *Studies in Conservation* Vol. 39 No. 1, 1994, pp. 39–44. [ECU-687]

Sharma, V.C., Uma Shankar Lal, and M.V. Nair. "Zinc Dust Treatment—an Effective Method for the Control of Bronze Disease on Excavated Objects." *Studies in Conservation* Vol. 40 No. 2, 1995, pp.110–119. [ECU-688]

Spriggs, J.A. "Treating the Coppergate Structures—A Quest for Antishrink Efficiency." *Archaeological Papers from York* M.W. Barley, P.V. Addyman and V.E. Black, eds. (1984): 78–86.

Suguwara, H. and Hideaki Ebiko. "Dezincification of Brass." *Corrosion Science* 7 (167): 513–523. [ECU-0136]

Sweeny, Louise. "Winning the Pigeon Battle of Washington." *Christian Science Monitor* (1991): 13. [ECU-366]

Taylor, R.J. and Ian D. MacLeod. "Corrosion of Bronzes on Shipwrecks: A Comparison of Corrosion Rates Deduced from Shipwreck Material and from Electrochemical Methods." *Corrosion* 41:2 (1985): 100–104. [ECU-144]

Trentelman, K.L. Stodulski, D. Scott, M. Back, D. Strahan, A.R. Drews, A.O.'Neill, H. Weber, A.E. Chen, and S.J. Garrett. "The Characterization of a new Pale Blue Corrosion Product Found on Copper Alloy Artifacts." *Studies in Conservation* Vol. 47 No. 4, 2002, pp. 217–227. [ECU-689]

Uminski, Maciej and Viviana Guidetti. "The Removal of Chloride Ions from Artificially Corroded Bronze Plates." *Studies in Conservation* Vol. 40 No. 4, 1995, pp. 274–278. [ECU-690]

Walker, R. "Benzotriazole as a Corrosion Inhibitor for Immersed Copper." *Corrosion* 29 (1973): 290–296. [ECU-145]

—. "Benzotriazole as a Corrosion Inhibitor," *Metal Finishing* 71:9 (1973): 63–66. [ECU-394]

—. "The Role of Benzotriazole in the Preservation of Copper Based Antiquities." *The Conservation and Restoration of Metals* (Proceedings of the Edinburgh Symposium, Scottish Society for Conservation and Restoration, Edinburgh) (1979): 40–49.

Watanabe, Masamitsu, and Yasohiro Higashiand Tohru Tavaka. "Difference Between Corrosion Products Formed on Copper Exposed in Tokyo in Summer and Winter." *Corrosion Science* Vol. 45 No. 7, 2003, pp. 1439–1453. [ECU-691]

Weisser, T.D. "The De-Alloying of Copper Alloys." *Conservation in Archaeology and the Applied Arts* N. Brumellie and P. Smith, eds.(Proceedings of the 1975 Stockholm Congress, International Institute for Conservation of Historic and Artistic Works, London) (1976): 207–214.

Zhang, Xiaomei. "An Unusual Corrosion Product, Pyromorphite, from a Bronze An: a Technical Note." *Studies in Conservation* Vol. 47 No. 1, 2002, pp. 76–79. [ECU-692]

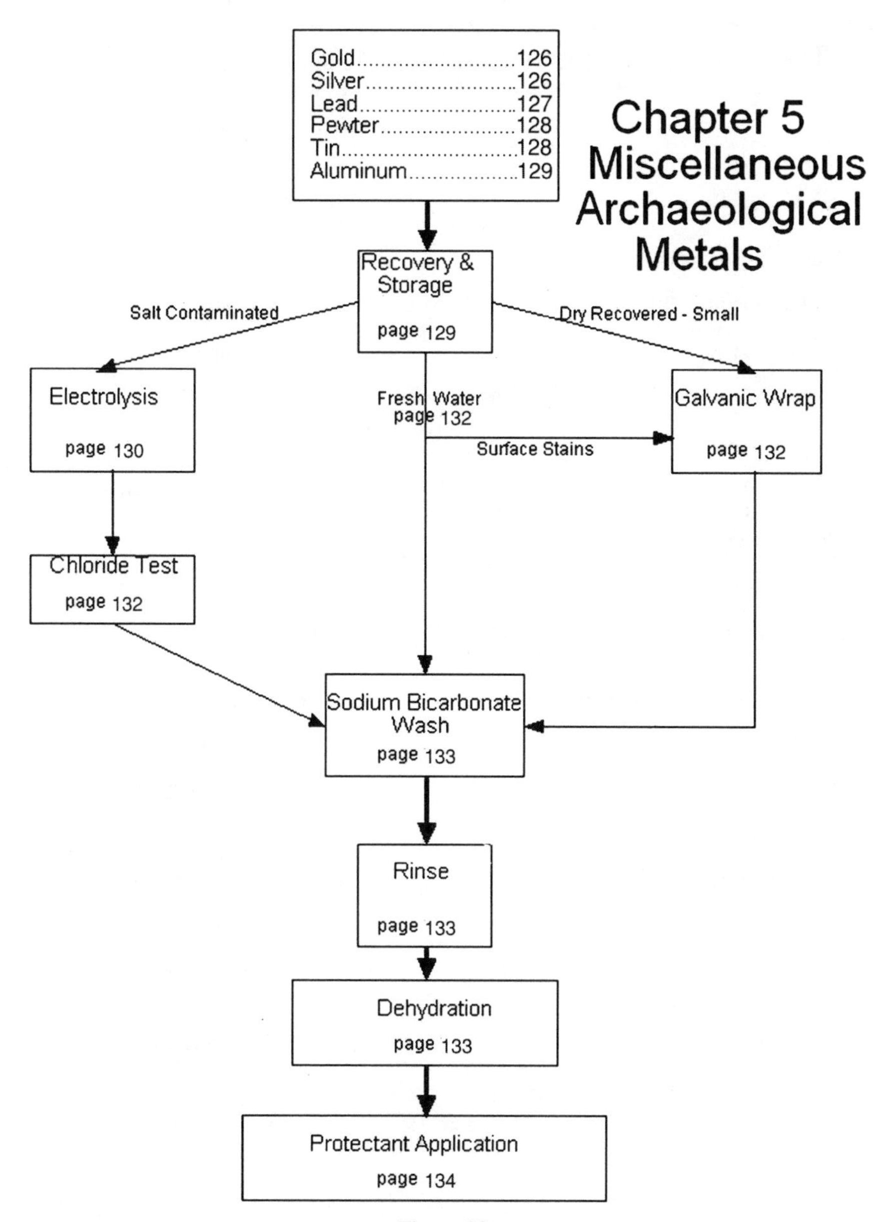
Gold....126
Silver....126
Lead....127
Pewter....128
Tin....128
Aluminum....129
Chapter 5
Miscellaneous Archaeological Metals
Recovery & Storage
page 129
Salt Contaminated
Dry Recovered - Small
Electrolysis
page 130
Fresh Water
page 132
Surface Stains
Galvanic Wrap
page 132
Chloride Test
page 132
Sodium Bicarbonate Wash
page 133
Rinse
page 133
Dehydration
page 133
Protectant Application
page 134

Figure 30.

Treatments for Miscellaneous Metals and Alloys

Represented in Conservation Literature

Alkaline Sulfite Reduction..Toxic Factors

Alkaline Sodium Dithionite..Toxic Factors

Amonia (Liquid)..Toxic Factors

Amonium Sulfate ($(NH_4)_2SO_4$ - M ascagnite)...............Toxic Factors

Amonia Amonium Sulphate...Toxic Factors

EDTA..Surface Treatment

Electrolytic Reduction
- High Current Density
- Moderate Current Density
- Low Current Density
- Low amperage/ Light Electrolyte..........................Recommended (page 130)
- Higher Amperage/ Light Electrolyte

Galvanic Wrap (Electrochemical Cleaning)......................Recommended (page 132)

Phosphoric Acid...Surface Treatment

Mechanical Surface Cleaning..Surface Treatment

Orthophosphoric Acid..Surface Treatment

Sequestering Agents.. ....Surface Treatment

Sodium Carbonate Wash..Slow, blocked by corrosion

Stripping Acids...Surface Treatment
- Oxalic
- Thioglycollic
- Citric...Will clean concretion
  - Modalene
  - Biox

Figure 31.

Chapter 5

Miscellaneous Archaeological Metals

Au, Ag, Pb, Pewter, Sn, Al

MISCELLANEOUS METALS—THEORY

Each of these metals, gold, silver, lead, pewter, tin, and aluminum, has its own history, morphology, chemistry, and physical traits. Among these, only pewter is an alloy and not a pure metal but as it is important historically and archaeologically, it deserves a place in this pantheon of pure elements. In an ideal situation, each of these materials deserves a chapter in its own right, but in the interest of keeping this manual organized and of a usable length, they will be gathered together in this chapter.

In practical application, with the exception of lead, most of these metals will be rare finds in an archaeological site. After all, humans have gone to extraordinary lengths to find and recover gold and silver and though these noble metals degrade slowly and last well in the archaeological record, they are not easily left behind or abandoned. In contrast, tin corrodes back to its mineralized versions quite easily while aluminum enters the archaeological record so late (only late 19th and early 20th centuries) that its greatest impact will be felt in the new field of aircraft archaeology.

Yet since they are all metals they share common traits. All metals corrode and this group is no exception. We have seen from the previous chapters concerning iron and copper that corrosion in metals is the act of a metal oxidizing to a more stable state. This is usually an oxidized or sulfate dominated mineral version of a metal, close if not identical to the ore from which the metal was refined. We have also seen that galvanic corrosion between two dissimilar metals in electrical contact is the donation of electrons from the metal that has a lower corrosion potential to the metal that has a higher corrosion potential. Some in this group, like gold and silver, are considered noble metals and will generally be protected in any galvanic linkage with another metal. So each of these metals will be discussed individually, but not in the same detail given to the more common archaeological finds of iron and copper.

Gold (Au)

Gold is a rare and stable heavy metal, known for its beautiful shine. Gold is soft, heavy, and ductile. It can be found in archaeological sites in ornaments, plating on ornaments, or in solid form in coins, and as bullion. Remarkably, it is relatively unaffected by burial or immersion in water and acquires very little oxidized surface patina so it tends not to tarnish. Like copper it can be found in nature in a relatively pure form. Gold is not poinsonous to living organisms, as are most heavy metals such as copper, lead, arsenic, cadmium and others, but since it does not corrode, little concretion can form on it. Gold generally remains unscathed in the archaeological record, though as mentioned, it is a rare find.

Silver (Ag)

Silver is another of the more noble metals. It is highly malleable and can be worked into ornaments or coins, bullion bars, or such decorative utilitarian items as candlestick holders, and flat ware. Silver is an oddity among the heavy metals with gold, in that it is non-toxic and its corrosion residue does not form an electrically passive surface layer in the earth or underwater. Silver is highly affected by penetrating anions and is rarely found in excellent condition underwater because of the amount of chloride anions that are available.

Silver invariably contains a high copper content that may partially passivate silver concretions, making them more protective of the artifact within. Dry finds

Figure 32. Silver Plated Cigarette Lighter manufactured by R. Wallace and Sons Co. The engraving only came to light during conservation. Photograph by Chris Valvano.

from land excavations tend to be in good condition with adherent dirt but not much concretion.

Aerobically produced silver concretions from sea sites (AgCl, cerargyrite—silver chloride), are usually dark gray or black, but can have some green color tinting due to copper corrosion products. Grey seabed anaerobic silver concretions (silver sulphide Ag2S argentitie, acanthite, and silver chloride/bromide) are surprisingly lighter in color than the aerobically formed concretions (Pearson, 1987; 92). The surface texture of these concretions is a rough gray matrix, but if the object were to be sectioned it would show multicolored layers. A typical layered coin might reveal an inner core of metallic silver surrounded by gray silver sulfide and on this a dark silver chloride layer. The chloride layer is the original surface of the coin and will contain all of the mint details. It is the surface that should be conserved rather than cleaning the artifact to the inner silver core. Over these layers is an outer rough concretion of gray silver sulfide formed when the coins lay buried in the sea bottom and were subject to anaerobic corrosion.

Experience has shown that recovered silver should never be allowed to dry or the outer concretion will adhere strongly to the original surface layer. On deconcreting the artifact, the archaeological surface will be damaged or destroyed.

Lead (Pb)

Lead is an easily worked heavy metal with a very low melting point. Lead can be alloyed with a lower percentage of tin (up to 30%) to produce solder. It is also used in pure form to produce pipes and some ship fittings such as patches, calking, scupper liners and hawser pipes. Lead shot is used for bullets, gaming dice, and fishing net weights among other items. Lead sheathing was also used in ancient times to combat hull fouling in wooden ships.

Lead is generally found in good condition on both land and underwater sites. Aerobic lead corrosion forms an electrically passive outer layer of lead sulfate ($PbSO_4$—Anglesite). Anaerobic lead does corrode faster than aerobic lead and the mineralized version is dominated by the sulfide ion (PbS—Galena).

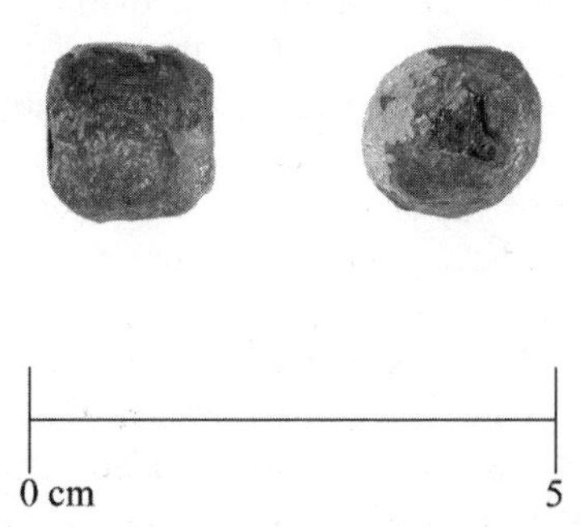

Figure 33. Lead gaming piece part way into its galvanic wrap conservation. On completion the concretions will have completely disappeared. Photograph by Chris Valvano.

Pewter

Pewter is historically an alloy of tin with varying amounts of antimony, copper, and lead. Some pre-18th century pewter demonstrates a lead content up to 50%. It became a popular kitchenware after medieval times and could be formed into such utilitarian items as plates, mugs, and utensils. By the mid 18th century both the health and tarnishing drawbacks of leaded pewter helped popularize Britannia metal, a non-leaded form of pewter consisting of tin, antimony, and copper.

Pewter has been found in good condition on both land sites and underwater, but its high tin content makes it a good candidate for selective galvanic corrosion in most underwater sites, similar to dezincified and destanified copper alloys. Tin's low corrosion potential (E corr) allows it to readily donate electrons to its alloying metals leaving tin ions free to migrate to a non-passive tin concretion. Logically then, tin with a high lead content will survive archaeological internment in the best condition with lead sulfite offering some electrical resistance to the concretion. Antimony and copper will also offer some corrosion resistance to pewter alloys but most pewter should be treated as though it were extremely delicate and brittle from selective galvanic corrosion.

Pewter concretions will be mainly formed of stannous (tin) oxide. Concretion should be left in place to avoid damage to the brittle artifact. Electrolytic reduction will easily exfoliate the concretion layer but it is unclear if it will reduce tin back to its metallic state.

Tin (Sn)

Tin may survive dry site internment better than it can on the sea floor. Even large ingots corrode on the seabed over time leaving a rough concreted corrosion product in the general shape of the original artifact. This concretion of stannous oxide consists of a crumbly gray matix. Tin has a lower E corr than copper and often selectively corrodes from bronze and pewter. It does, however, have a higher E corr than iron. This makes tin an ideal coating for iron and steel in the canning industry. Tin cans in the 19th century are actually made of alloys of iron and later steel that are coated on both sides in tin. Since the tin is galvanically protected from corrosion by the electrons given off by the iron, it became a cost-effective protective coating for cans. The non corroding tin coating keeps the iron from leaching into the food and even protects the iron in the can from some of the acidic foods that are stored inside. 19th century tin cans (and metal buckets) are held together at the seam by solder, an alloy of lead and tin.

Below is a table of common archaeolgically recovered metals in order of their corrosion potentials (E corr) high to low. When an object made of one of the metals listed below comes into contact with an object above it in the table it will donate electrons and corrode. When an artifact made of one of the metals listed below comes into contact with a metal below it in the table, it will receive electrons from the lower listed metal object and be reduced.

Galvanic Metals Table (Most Noble at top left)

Gold	Lead
Silver	Pewter (leaded 50%)
Copper	Wrought Iron
Nickel	Cast Iron
Low Bronze	Pig Iron
High Bronze	Steel
Low Brass	Aluminum
High Brass	Zinc
Muntz Metal	Magnesium Alloys
Tin	Magnesium

Aluminum (Al)

Aluminum was not commercially produced until 1886 and does not, therefore, significantly impact archaeological analysis and should not generally be found on early archaeological sites. Aluminum conservation will, nonetheless, become a popular topic in the fledgling field of aviation archaeology. The metal's lightweight made it ideal for the construction of airplanes and airships.

Aluminum alloys usually contain a small amount of copper that will act as a cathodic surface to the dissolution of some of the surrounding aluminum (Pearson, 1987; 247). But aluminum oxide is generally electrically passive, giving it good corrosion resistance on land or in fresh water. Aluminum recovered from saltwater sites, however, can have a great deal of corrosion and exhibits some chloride ion infiltration, though not to the extent of other metals (Pearson, 1987; 247). Aluminum rarely is concreted although some aluminum magnesium alloys are hard enough to exhibit standard calcium carbonate concretions. Apparently the aluminum oxides are easily dislodged from standard aluminum in a high energy environment. Corrosion accelerates if the passive oxides cannot protect the metal benieth. In addition, the newly exposed surface makes life difficult for fouling organisms that attempt to fasten themselves to the artifact and no concretion forms to protect the aluminum. These factors lead to extensive aluminum corrosion in high-energy environments. Rinsed aluminum/magnesium alloy can loose up to 6% of its weight in contaminants, chlorides, and pitting residue. Pitting is generally extensive in aluminum/magnesium alloys and there is no evidence that that these alloys reduce back to a metallic form in electrolysis.

RECOVERY AND STORAGE—METHODOLOGY

As with all of the other metals discussed thus far, gold, silver, lead, tin, and pewter, will all benefit from the standard recovery techniques advocated in this manual. Immediately on recovery an artifact should be wrapped in aluminum foil and submerged in a container of 2% to 5% sodium carbonate or sodium bicarbonate (baking soda). As with other recovered metals the solution should be allowed to enter the foil wrapping. The foil will begin the reduction process while the artifacts are in transit or storage. For some artifacts recovered from fresh water sites this may be the only treatment necessary to insure their continued stabilization. Obviously from the previous discussions, aluminum artifacts need not be wrapped in aluminum foil, as aluminum foil will not galvanically couple with an aluminum artifact.

It is especially imperative that silver concretions not be allowed to dry as the archaeological surface may adhere to the concretion and be lost in conservation. As with copper and iron, no dissimilar metal objects should be allowed to touch in storage (see galvanic chart), though this is somewhat mitigated by the protection afforded by the aluminum foil wrap.

Artifacts recovered from saltwater environments, or that contain corrosion products and or concretions, should undergo electrolytic reduction. Objects recovered from fresh water can often bypass electrolysis and go directly to the sodium bicarbonate wash. Artifacts recovered from a dry site and fresh water will benefit from a galvanic wrap treatment.

ELECTROLYSIS—METHODOLOGY

Electrolytic reduction as described for iron will work well for most of these miscellaneous metals provided, of course, there is a core of sound metal to make the electrolytic connection. This is a large presumption, and the process will actually become destructive if there is no sound metal within the artifact. Electrolytic action will remove the concretion surrounding an artifact. This process can be destructive if the archaeological surface needs to be saved but is not metallic, as is often the case with silver coins and ornaments. Therefore, included below are hints for each of these metals undergoing electrolysis.

Gold

Does not usually need electrolysis, a galvanic wrap is much simpler and would reduce the artifact as needed.

Silver

Electrolysis works well on most silver artifacts, the one exception being the reduction of coins to their metallic cores. The archaeological surface of many coins may be a dark gray or black silver chloride layer inside the concretion that will be lost in a full electrolysis of the core. This procedure will exfoliate the outer layers including the silver chloride layer that contains the surface of the coin including the mint mark and date. Though there are techniques that can be used to gently clean the coins to their archaeological surface (unless the concretion has dried) these techniques should not normally be attempted at a basic laboratory. In the case of layered coins a consultant should be called.

Otherwise, electrolytic reduction of silver can be accomplished in the manner described for iron and copper, and will also be described at the end of this section.

Lead

There are no problems associated with reducing lead, electrolysis works well on this metal and will remove a majority of the corrosion products and concretion. Mild steel anodes are used so as not to contaminate the electrolyte with poisonous lead ions.

Pewter

Most pewters can also be reduced through electrolysis. Set up is the same as it is with iron and copper using tin or mild steel anodes.

Tin

Tin recovered from a dry site may be in good enough shape to attempt electroysis. Tin recovered from an underwater site may be too extensively corroded to be able to ground the core of the artifact. In this case the artifact can literally break up or become a sludge on the bottom of the electrolytic tank. If there is a question concerning the metal's soundness a consultant should be called.

Aluminum

Aluminum electrolytic rinse is problematic for heavily chloride anion contaminated objects. It works well for dry recovered artifacts. Electrolysis on seabed recovered artifacts exposes vast amounts of pitting corrosion. This corrosion has taken place on the artifact and remains hidden until electrolysis. At this point heavily chloride contaminated aluminum should not undergo electrolysis. To date, chemical rinses such as the various alkaline washes offer a better alternative than electrolysis for aluminum.

The electrolytic reduction for these metals is carried out using low amperage reduction described for iron. Mild steel anodes can be used for each of these metals, though the reader is reminded that magnetite will plate out on the artifacts and they may have to be scrubbed free of this black residue with baking soda paste and a nylon brush. This black residue is particularly unnatural on aluminum and silver. The conservator, therefore, may wish to expend some effort locating anodes of the same material as the artifact—disposal of heavy metal electrolytes should involve allowing the water to evaporate and dealing with the residue in a responsible manner. It should also be noted again that most electrolytic set ups fail because the connections do not allow a continuous circuit. A rule of thumb is that no connections should be made until and unless bare shiny metal on the artifact and anodes are in contact with the alligator clip connectors. An electrician's multi-meter set to test resistance should be used to test all electrical pathways and connections before the electrolyte is added and the current applied to the artifact.

Sodium carbonate is a good non-toxic electrolyte that can be used for all of these various metals. Solution percentages of .25% to .5% will control the current amperage of the DC power source to 7 amps or less.

CHLORIDE TEST—METHODOLOGY

Chloride testing of the electrolytic solution of smaller artifacts that are manufactured from the listed miscellaneous metals can be confusing and is not recommended for all artifacts. Often these metals, due to their small size and lack of porosity, will not contain enough chlorides, even if they are heavily contaminated, to raise the chloride ion percentage in the electrolyte solution to meaningful or measurable levels (see iron chloride measurement discussion). General time frames based on artifact size are actually more useful than attempting a graph of chloride ion concentration versus time for every artifact. Small artifacts of pin, or buckle size receive one to two weeks of electrolysis. Larger artifacts, of spike or plate size, should undergo reduction for about a month. Renewed corrosion in storage (very rare) is the best indicator that the artifact has not been thoroughly rinsed.

FRESH WATER RECOVERED METALLIC ARTIFACTS—METHODOLOGY

Fresh water recovered artifacts of these assorted metals will invariably be in remarkable condition. Experience has demonstrated that these artifacts will seldom show much corrosion and will almost never have any concretion. If there is concretion or an oxidized patina they should be galvanically wrapped. If these artifacts show no patina or oxidation they should be sent to the sodium bicarbonate wash phase.

GALVANIC WRAP—METHODOLOGY

Small artifacts, of fastener or shoe buckle size, can be galvanically wrapped. This is essentially no diffenent than their recovery treatment. They will be mechanically cleaned and placed in an aluminum foil pouch that contains sodium carbonate electrolyte in the 2% to 5% range. Gold and silver may actually become clean and shiny faster from a citric acid electrolyte rather than sodium carbonate, but both perform well. The foil pouch should be wrapped as tightly about the artifact as is possible without doing harm to the artifact. Next the artifact is placed into a beaker of the same electrolyte as that in the pouch—within days the aluminum foil will corrode as it reduces the artifact in the pouch and begins to leak.

The wrap should be opened and examined every three days or so. After three days the artifact should show signs of reduction, its surface patina should begin to look like the elemental metal rather than a dull oxidized form. Adherent dirt and sand will have fallen away and any small concretion should dissolve. It should be remembered here that many artifacts such as silver bells are manufactured as a composite. The bells may have iron clappers and the clappers will have corroded galvanically to protect the silver. A galvanic wrap will protect and reduce both metals, though it is possible the iron has completely oxidized and cannot be saved.

SODIUM BICARBONATE WASH—METHODOLOGY

As with iron and copper alloys the miscellaneous metal artifact should undergo a thorough rinse and scrub using soft nylon bristle toothbrushes and a paste of water and sodium bicarbonate. If the artifact is in robust condition it may be boiled in sodium carbonate, sodium bicarbonate, or sodium sesquicarbonate for up to several hours. This should clean and release any anions that were not rinsed in the galvanic wrap, since it is a rather passive reduction and rinse process.

FINAL RINSE—METHODOLOGY

The final rinse should remove all of the residue of the sodium bicarbonate wash. It is accomplished with a thorough soak in distilled water. Agitation will speed the rinse considerably, but a static rinse should take several hours with one change of rinse water.

DEHYDRATION—METHODOLOGY

Solvent dehydration is recommended for all of these miscellaneous metals except tin and aluminum that can withstand oven dehydration. Composite artifacts or those with plating, guilding, or any other coating should also be solvent dried. Artifacts should be taken directly from the rinse and placed in three successive baths of alcohol or acetone (see copper dehydration). Each bath should last one hour. Used solvent will suffice for the first two baths but the final bath should be fresh solvent. As always the solvents are highly flammable and should be covered at all times and never used near an ignition source. All manipulation of artifacts after the wash phase should be done with gloves or utensils as salt on the fingertips will damage artifacts over time.

Oven drying in a 325 degree F (163 degrees C) oven for two days does not injure aluminum or tin. It should be remembered, however, that some artifacts such as tin cans are actually composite artifacts and should be solvent dried.

Artifacts should not be removed from their dehydration treatments until the conservator is ready to apply a protectant, or unless a desiccator is available. A desiccator is simply an airtight container partially filled with dessicant (a salt that absorbs atmospheric humidity).

PROTECTANT APPLICATION—METHODOLOGY

At times it may not be deemed necessary or appropriate to apply a protectant to an artifact. Gold will seldom need protection from humidity and silver protectant application depends on it use after conservation. In these cases, the artifacts should be stored in dessicant chambers or in a storeroom with very low relative humidity (preferably 40% or less).

If on the other hand, artifacts are to be handled or displayed or will be subject to uncontrolled environments, they may need a surface coating to protect them from moisture absorption. Artifacts that underwent solvent dehydration (gold, silver, lead, pewter) can be coated with shellac. Shellac is a hard impermeable coating that can be applied to artifacts that are heat sensitive. Shellac can be removed if necessary with alcohol and its' surface luster is easily changed from glossy to dull with a buffing of steel wool.

Aluminum and tin can be microcrystalline wax coated using the same methods prescribed for copper (page 119) and iron (page 95–96). The micro-wax will dehydrate and seal the artifact against further humidity. The objects are placed into the melted wax and allowed to stay two hours beyond their last effervescence. The wax is then turned to low and cooled slightly before the artifacts are removed to drip dry and cool. After cooling, excess wax can be removed with a knife or smoothed with a fingertip.

CONCLUSION

The metals discussed in this chapter will make up only a small portion of the artifacts recovered on any archaeological site, hence their proportionately smaller representation in this manual. Nonetheless, the intrinsic value of these objects may outweigh their numerical representation in the archeological record. Silver, gold, and pewter represent both ornamental and utilitarian objects, bringing perhaps a better balance to site interpretation than is possible with purely utilitarian objects of iron and copper. It is up to the conservator then, to make sure that these objects remain stable from recovery to storage, in order that they can be reinterpreted at any time.

Fortunately, most noble metals behave in a predictable manner and are some of the most stable elements found in nature. These will require only very basic treatment. Others, like pewter, are alloys of both stable and unstable metals, making their durability and integrity less predictable. Nonetheless, their value to the archaeological interpretation of a site is almost always enhanced with the detail brought about by the conservation process.

CHAPTER 5: MISCELLANEOUS ARCHAEOLOGICAL METALS

"Aqueous Corrosion of Tin-Bronze and Inhibition by Benzotriazole." *Corrosion* Vol. 56 No. 12, 2000, pp. 1211–1218. [ECU-695]

Britton, S.C. *The Corrosion Resistance of Tin and Tin Alloys*. Greenford, U.K.: Tin Research Institute, 1952.

Caley, E.R. "Coatings and Encrustations on Lead Objects from the *Agora* and the Method Used for their Removal." *Studies in Conservation* 2/2 (1955): 49–54.

Carlin, Worth and Donald H. Keith. "On the Treatment of Pewter Plates from the Wreck of *LaBelle*, 1686." *International Journal of Nautical Archaeology* Vol. 26 No. 1, 1997, pp. 65–74. [ECU-693]

Carradice, I.A. and S.A. Campbell. "The Conservation of Lead Communion Tokens by Potentiostatic Reduction." *Studies in Conservation* Vol. 39 No. 2, 1994, pp. 100–106. [ECU-694]

"Cavitation Corrosion Behavior of Cast Nickel-Aluminum Bronze in Sea Water." *Corrosion* Vol. 51 No. 5, 1995, pp. 331–342. [ECU-696]

"Corrosion of Steel in Tropical Sea Water." *British Corrosion Journal* Vol. 29 No. 3, 1994, pp. 233–236. [ECU-697]

Cronyn, J.M. *The Elements of Archaeological Conservation*. London: Routledge, 1990. [ECU-350]

Davis, Mary, Fraser Hunter, and Alec Livingstone. "The Corrosion, Conservation, and Analysis of a Lead and Cannel Coal Necklace from the Early Bronze Age." *Studies in Conservation* Vol. 40 No. 4, 1995, pp. 257–264. [ECU-698]

Dean, J.S. "the Medea Floats Again: Steel Hull Preservation Using Foam Core and FRB." *American Neptune* Vol. 60 No. 2, 2000, pp. 131–147. [ECU-699]

Degrigny, C. and R. LeGall. "Conservation of Ancient Lead Artifacts Corroded in Organic Acid Environments: Electrolytic Stabilization/Consolidation." *SIC* Vol. 44 No. 3, 1999, pp. 157–169. [ECU-700]

Gardinour, C.P. and R.E. Melchers. "Corrosion of Mild Steel by Coal and Iron Ore." *Corrosion Science* Vol. 44 No. 12, 2002.

—. "Corrosion of Mild Steel in Porous Media." *Corrosion Science* Vol. 44 No. 11, 2002. [ECU-669]

Gottlieb, Adam. "Chemistry and Conservation of Platinum and Palladium Photographs." JAIC Vol. 34 No. 1, 1995. [ECU-729]

Hamilton, Donny L. *Basic Methods of Conserving Underwater Archaeological Material Culture*. Washington D.C.: U.S. Department of Defense Legacy Resource Management Program, January 1996. [ECU-610]

Hoare, W.E., and E.S. Hedges. *Tinplate*. London: Edward Arnold, 1945.

Holm, S.I. "Silver." (Master's Thesis. Institute of Archaeology, University of London) (1969).

Kato, M., M. Koiwai, and J. Kuwano. "The Aluminum Ion as a Corrosion Inhibitor for Iron in Water." *Corrosion Science* 19:11 (1979): 937–947. [ECU-172]

Kaufmam, Henry. *The American Pewterer: His Techniques and His Products*. New Jersey: Thomas Nelson, 1970.

Johnson, Colin, Kerry Head, and Lorna Green. "The Conservation of a Polychrome Egyptian Coffin." *Studies in Conservation* Vol. 40 No. 2, 1995, pp. 73–81. [ECU-701]

Lane, Hannah. "Reduction of Lead." *Conservation in Archaeology and the Applied Arts* (Proceedings of the 1975 Stockholm Congress, N. Brommelie and P. Smith, eds., International Institute for the Conservation of Historic and Artistic Works, London) (1975): 215–217.

—. "Some Comparisons of Lead Conservation Methods, Including Consolidative Reduction." *Conservation and Restoration of Metals* (Proceedings of the Edinburgh Symposium, Scottish Society for Conservation and Restoration, Edinburgh) (1979): 50–60.

Lee, S. and R.W. Staehle. "Absorption of Gold." *Corrosion* Vol. 52 No. 1, 1996, pp. 843–852. [ECU-702]

Leidheiser, Henry. *The Corrosion of Copper, Tin, and Their Alloys*. New York: John Wiley, 1971.

MacLeod, Donald Ian. "Stabilization of Corroded Aluminum." *Studies in Conservation* 28 (1983): 1–7. [ECU-119]

MacLeod, Ian Donald and Neil A. North. "Conservation of Corroded Silver." *Studies in Conservation* 24 (1979): 165–170. [ECU-115]

Marshall. A., Piercy , R., and N.A. Hampson. "The Electrochemical Behaviour of Lead/Tin Alloys—Part I: Studies in Nitrate Electrolytes." *Corrosion Science* 15 (1975): 23–34. [ECU-138]

Mor, E.D. and A.M. Beccaria. "Inhibitory Action of Acryonitrile on the Corrosion of Zinc in Seawater." *British Corrosion Journal* 8:1 (1973): 25–27. [ECU-146]

Organ, Robert M. "The Reclamation of the Wholly Mineralized Silver in the Ur Lyre." *Application of Science in Examination of Works of Art* (Proceedings of the Seminar Musem of Fine Arts, Boston) (1967).

Pearson, Colin ed. *Conservation of Marine Archaeological Objects*, London: Butterworths Series in Conservation and Museology, 1987. [ECU-6]

Plenderleith, H.J., and A.E.A. Werner. *The Conservation of Antiquities and Works of Art.* Second edition. London: Oxford University Press, 1971. [ECU-8]

Richardson, J.A. and G.C. Wood. "A Study of the Pitting of Aluminum by Electron Microscopy." *Corrosion Science* 10 (1970): 313–329. [ECU-194]

Rodgers, Bradley A. *The East Carolina University Conservators Cookbook: A Methodological Approach to the Conservation of Water Soaked Artifacts*. Herbert Pascal Memorial Fund Publication, East Carolina University, Program in Maritime History and Underwater Research, 1992. [ECU-402]

—. *Conservation of Water Soaked Materials Bibliography*. 3rd ed. Herbert Pascal Memorial Fund Publication, East Carolina University, Program in Maritime History and Underwater Research, 1992.

"Sacrificial Anode Cathodic Polarization of Steel in Seawater: Part 1—A Novel Experimental and Anaylsis Methodology." *Corrosion* Vol. 52 No. 6, 1996, pp. 419–427. [ECU-703]

Sease, Catherine, Lyndsie S. Selwyn, Susana Zubiate, David F. Bowers and David E. Atkins. "Problems with Coated Silver: Whisker Formation and Possible Filiform Corrosion." *Studies in Conservation* Vol. 42 No. 1, 1997, pp. 1–10. [ECU-706]

Scott, David A. "The Deterioration of Gold Alloys and Some Aspects of their Conservation." *Studies in Conservation* 28 (1983): 194–203. [ECU-391]

Singley, Katherine, R. *The Conservation of Artifacts form Freshwater Environments*. South Haven, Michigan: Lake Michigan Maritime Museum, 1988. [ECU-10]

Slack, C.G. "Technical Notes on the Cleaning and Reproduction of Silver Coins." (Irish Archaeological Research Forum) 1:1 (1974): 52–58.

Sramek, J., T. Jakobsen, and Jiri B. Pelikan. "Corrosion and Conservation of a Silver Visceral Vessel from the Beginning of the Seventeenth Century." *Studies in Conservation* 23 (1978): 114–117. [ECU-177]

—. "The Corrosion of Lead and Tin: Before and After Excavation." *Lead and Tin Studies in Conservation and Technology* (Occasional Papers No. 3, G. Miles and S. Pollard, eds., United Kingdom Institute for Conservation, London) (1985): 15–26.

—. "The Behavior of Lead as a Corrosion Resistant Medium Undersea and in Soils." *Journal of Archaeological Science* 10:4 (1983): 397–409.

"Technical Note: Investigation into Tarnishing of Pewter Artefacts Recovered from the *Mary Rose*." *British Corrosion Journal* Vol. 32 No. 3, 1997, pp. 223–224. [ECU-707]

Tennant, N, Tate J and Cannon, L. "Corrosion of Lead Artifacts in Wooden Storage Cabinets." *Scottish Society for Conservation and Restoration* Vol. 14 No. 1, 1993. [ECU-787]

Tetreault, Jean, Jane Sirois, and Eugenie Stamatopoulou. "Studies of Lead Corrosion in Acetic Acid Environments." *Studies in Conservation* Vol. 43 No. 1, 1998, pp. 17–32. [ECU-708]

Walker, Robert and Alexandra Hildred. "Manufacture and Corrosion of Lead Shot from the Flagship *Mary Rose*." *Studies in Conservation* Vol. 45 No. 4, 2000, pp. 217–225. [ECU-709]

Watson, Jaqui. "Conservation of Lead and Lead Alloys using EDTA Solutions." *Lead and Tin Studies in Conservation and Technology* (Occasional Papers No. 3, G. Miles and S. Poilard, eds., United Kingdom Institute for Conservation, London) (1985): 15–26.

Wise, Edmund M., ed. *Gold—Recovery, Properties, and Applications*. Princeton: Van Nostrand, 1964.

Chapter 6 Archaeological Ceramic, Glass, and Stone

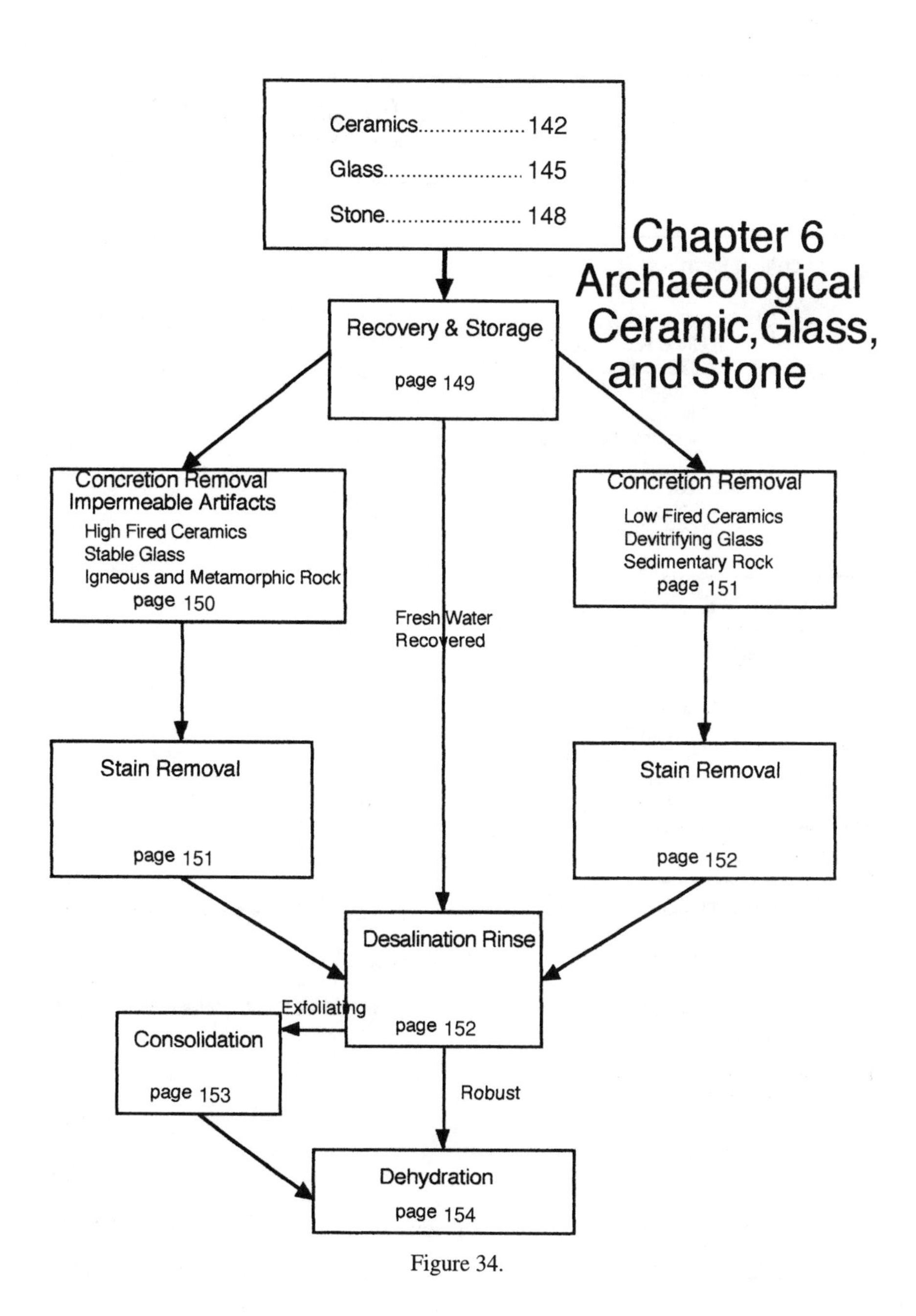

Figure 34.

Treatments for Ceramic, Glass & Stone
Represented in Conservation Literature

Concretion Removal

Chemical.......................................Acids -Robust Artifacts
- Citric Acid.......................................150
- Oxalic Acid
- Phosphoric Acid
- Hydrochloric Acid
- Sodium Hexameaphosphate.......................................150

Mechanical
- Scalpel
- Dental Picks

Stain Removal

Chemical
- Citric Acid.......................................151
- Oxalic Acid
- Phosphoric Acid
- Hydrocloric Acid
- Hydroflouric Acid
- Di, Tri, & Tretrasodium Salt of EDTA.......................................152

Desalination Wash

Static Desalination Rinse.......................................Delicate Artifacts - 153

Agitation Rinse.......................................152

Flow Through Rinse.......................................Needs Vast Quantities Fresh H_2O

Consolidation

B -72.......................................153

PVA.......................................153

Silicon Oil.......................................Not Reversible

Penetrating Epoxy.......................................Not Reversible

Figure 35.

Chapter 6

Archaeological Ceramic, Glass, and Stone

CERAMIC, GLASS, AND STONE (LITHICS)—THEORY

Ceramic, glass, and stone are related for many reasons. First, they represent fabricated artifacts that are among the most durable produced by man. Their robust nature is often reflected in the fact that they represent a large portion of the remaining material culture found on archaeological sites. Storage ware, bricks, flatware, hollow ware, and bottles litter historic sites, while ceramic and knapped lithics (stone), projectile points, bi-face blades, and percussion flakes are some of the more diagnostic and long lasting artifacts found at prehistoric sites. Though artifacts fabricated of these materials may be broken, the material from which they were constructed is seemingly indestructible. The fact remains with most of these materials that entry into the archaeological record, through breakage, disaster, or loss, remains the most destructive event visited upon them.

Many ceramic, glass and stone, artifacts are also related by their fabrication and mineral content. Plenderleith and Werner consolidate them into the category of "Siliceous and Related Materials," for their silicate content (Plenderleith, 1971; 299–343). Most of the earth's crust is made of various silicates like silica (sand), clay, talc, feldspar, or other silicates in one form or another. Silicates remain strongly related even if their mineral content is not identical. Most of these materials, with the exception of sedimentary rock, were formed either naturally or by man, through application of great heat. This gives them certain characteristics in common and though they may not be formed of similar elements, they may be treated in conservation in a similar manner.

The durability of ceramics, glass, and stone, should make conservation of these materials a seemingly simple matter. However, the reality is that while ceramic, glass and lithics do offer the conservator some flexibility and for the most part are forgiving of rough handling, storage, and treatment, they are not a monolithic category any more than is glass or stone. Some of these materials are durable while others, depending on mineral content, manufacuring technique and fabrication temperature, can be quite fragile. In any case, recovery and removal from their stable earth bound or sea floor environment will begin a decomposition processes that must be mitigated.

It is important in this instance that the archaeologist/conservator be able to recognize the material they are working with. High fired stoneware and porcelain should be distinguished from terra cotta and earthenware. In the same way, limestone and marble should be distinguished from granite or basalt. It is beyond the scope of this work to relate ceramic, glass, and geologic taxonomy but its importance once again underscores the close relationship between archaeologist and the conservator. These professionals should be trained or make themselves cognizant of material taxonomy and typology as it forms the basis for all conservation work.

The overall durability of this class of material is reflected in the fact that there are far fewer research sources than there are for other materials, reflecting the squeaking wheel nature of unstable artifacts versus stable material, as well as today's economic engineering concerns over metals. But it also reflects the fact that although there are books and studies devoted to ceramics and glass, they cater mainly to collectors and art admirers. These sources are mainly concerned with the appearance and artistic presentation of an artifact, not necessarily with in depth material analysis or fabrication technique. This is where archaeology/conservation crosses into the fields of engineering and industrial technology. Geological studies abound concerning lithic taxonomy and mineral and rock formation, yet few if any are devoted to the study of how human interaction affects stone. The following brief summaries of the knowledge collected in the general works is designed to give the reader a basis and understanding of the conservation techniques chosen for ceramics, glass, and stone.

CERAMICS—THEORY

Pottery shards can normally be found in quantity in human occupation sites both historic and prehistoric, while underwater sites also abound with ceramic. Ships often carried large quantities of storage ware and kitchenware for trade. Ceramic and pottery first used around 10,000 before present, ushered in the neolithic age. Ceramics offer humans a means of storing collecting and transporting agricultural and mineralogical commodities that had not previously existed. Pottery's resistance to liquid and heat changed human cooking habits and went hand in hand with the development of agriculture and stored fermented beverages.

Pottery remains fairly basic throughout time. In its simplest terms it consists of shaping a vessel out of clay and baking it until it becomes a hard mass that is fairly impervious to re-hydration. Clay consists of weathered silicates in the 4 micron size range that generally take the color of whatever mineralized metal that lies in the soil. Iron oxides for example will create a red colored clay, while aluminum salts make for a white kaolin clay.

Early potters discovered that pure clay invariably cracks when it is fired. Pottery, therefore, needed an added temper to eliminate cracking. Virtually any substance can become a temper and archaeologically tempers can consist of sand, crushed shell, grass, manure, or ash. The temper can become a good identifying element of any ceramic.

Prehistoric pottery was invariably made by hand with the building of a three-dimensional shape. In vessels this may include various techniques of pinching and solid forming. Later more sophisticated techniques included the coil, slab or paddle methods. Each method is accomplished just as its name implies with the vessel built up of coiled rolled clay or of slabs pinched together at the seams. Molding clay in plaster molds and turning vessels on wheels are considered historic developments and require equipment beyond the human hand.

By historic times ceramics could be distinguished by the types of clays used, their mineral content, their glaze, the decorative pattern used, and their firing temperature. Broadly speaking ceramics can be divided into two groups, low-fired ware and high fired ware. This distinction is important in determining which conservation process should be used on a particular ceramic piece.

Low fired ware includes terra cotta fired at less than 1000 degrees centigrade and earthenware fired between 900 and 1200 degrees centigrade. Terra cotta can be very delicate and includes most prehistoric pottery. Historic earthenware is more robust having been fired at a higher temperature. Earthenware is red, yellow or buff colored depending on the mineral content. Though fired at higher temperature than terra cotta the pottery is not waterproof and must be coated in a glaze or slip to make it impermeable. But even so, a conservator should be aware that earthenware pottery can suffer from salt infiltration problems.

Archaeological ceramics are generally not found intact, this allows the infiltration of aqueous borne contaminants, such as salts, into the body of the pottery piece or shard. Stains and salt infiltration can occur on dry land sites as well as underwater. Dissolved salts can infiltrate ceramic bodies and crystallize when the ceramic piece dries. Since crystals expand they may force glaze off the ceramic or even flake pieces of the body. Stains are indicators of metallic salt intrusion, iron is typically orange and copper alloys are blue or green. Left in place the acids these stains represent will break down the body of a ceramic. Black sulfide organic stains can also penetrate the body of a piece of ceramic and will begin decomposition of the piece.

Earthenware recovered from the sea floor can be covered in calcarious deposits left by fouling organisms. Parrot fish and stone borers will also damage the surface of exposed ceramics. Parrot fish, in particular, seem to delight in leaving scrape marks in a tell tale tic-tac-toe cross hatching as they graze with their sharp teeth on the algae growing on the outer surface of the pottery.

The following is a reference list for the archaeologist/conservator of pottery considered low fired and the approximate temperatures in degrees C. at which they were produced.

High-fired ware includes stoneware fired between 1200 and 1300 degrees C. and china or hard paste porcelain fired between 1250 and 1450 degrees C. High-fired ware forms a hard impenetrable body that sinters or melts to produce a material impervious to water and salt contamination and, therefore, is a much easier material to conserve than low fired ware. True hard paste porcelain was not

Low-Fired Wares

Terra Cotta	
Mud Brick	Sun Dried
Terra Cotta (under-fired)	400–650
Terra Cotta	<1000
Earthenware	
China Glazed Creamware	1050
Creamware (Queensware)	1050
Delft (Dutch Imitation Porcelain)	1050
Faience (French Imitation Porcelain)	1050
Industrial Slipware	NA
Iron Stone	1200
Jackware	1150
Marble Slipware	NA
Agate Ware	NA
Mojolica (Italian imitation Porcelain)	1050
Pearl White or China Glaze (known as Pearlware)	1150
Potuguese Faience (Portuguese Imitation Porcelain)	1050
Red Bodied Earthenware	900
Slipware	1050
Sgraffito	NA
Spode	NA
Turnerware	NA
White Granite	NA
Whiteware	1200
Whitestoneware	NA

produced in Europe until Freidrich Bottgers of Meissen Germany first produced it in the early 18th century (Hunter, 2003: 137).

High-Fired Wares

Stoneware	
Salt Glaze Stoneware	1250
Gray Bodied Stoneware	1250
Jasper	NA
Rosso Antico	1250
White Salt Glaze Stoneware	1250
Porcelain	
Chinese Hard Paste	1250–1450
European Hard Paste	1250–1450

(Pearson, 1987: 100; Hunter, 2003)

High fired ceramics can have surface stains and are subject to calcarious concretion build up in an oceanic environment. But the stains and concretion will not penetrate the body of the ceramic as is likely with low-fired wares. Though some stoneware is described as having salt or white salt glaze, it is not a true glaze.

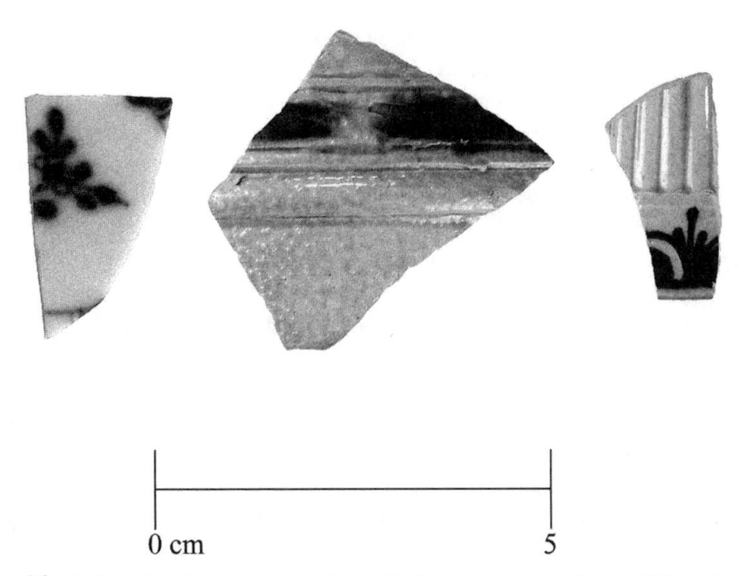

Figure 36. Left to right; Hard paste porcelain, salt glaze stoneware, and pearl white earthenware. Though somewhat similar in appearance, the impermeable porcelain and stoneware are stabilized differently than the lower firing temperature, earthenware. Photograph by Chris Valvano.

Salt glazing is a process whereby salt that is added to the kiln fire and evaporates in the high temperatures, to condense on the relatively cooler pottery. Salt glazing gives stoneware coloration and an orange peel texture. Salt glaze will not show up as a distinct layer on the stoneware as would a glaze. Surface cleaning and rinsing, are virtually all that high-fired wares need for stabilization.

True porcelain is the only pottery that is translucent. As with stoneware there will be no distinct glazing and no layering in cross section. First developed in China, porcelain was heavily sought after in Europe as a trade commodity. Europeans admired the fact that this pottery was translucent, thin walled and extremely durable, but they had no idea how it was made. An entire industry of imitation porcelain sprang up in Europe from the 15th century onward as Europeans produced soft paste earthenware such as Mojolica, Faience, and Delft well into the 18th century. During the 18th century Europeans finally deduced that true porcelain could be produced using kaolin clay (white aluminum salts) plus petuntse or Growan stone baked at higher than normal kiln temperatures (Hunter, 2003: 136).

GLASS—THEORY

Glass occurs in nature in the form of volcanic obsidian and melted silica sand from lightning strikes known as fulgerites. It was first produced artificially around 3000 BC as glaze for pottery. The first all glass vessels were produced much later about 1500 BC. and traded in ingots throughout the Mediterranean during the later Bronze Age. Early glass was often blue in color due to metallic copper salts in the glass mixture. Ancient Roman glassmakers propelled the art of glass making to new heights with development of the blowpipe shortly before the birth of Christ and window glass shortly after. The high art of the Roman glass blower was not again achieved in the West until the 19th century.

Unlike pottery, glass has to be worked at high temperature. Traditionally molten glass (the consistency of a stiff liquid) is produced in a vat in a hearth furnace and gathered on the end of a blowpipe or pontil. Blowing into the pontil creates a bubble in the gathered glass that can be shaped and worked into hollowware such as bottles and vases. Each time the glass worker cuts or shapes the glass it has to be inserted into the furnace and reheated. Additional gathers can add detail and ornamentation to the glass. Glass workers are true artisans that acquire more skill with each object produced.

Archaeologically, glass is a rare find in the New World until the 17th century. Bottles and other vessels were produced in both southern and northern Europe for a few centuries before this. Glass blowers could mass-produce bottles by blowing glass into heat resistant molds and then applying more glass to the neck after it was removed from the pontil. Plate glass does not come into general use until the end of the 17th century and even then it is expensive and fairly rare on most archaeological sites.

The main ingredient of artificially produced glass is silica (sand) heated to produce a vitreous mass. Melted silica's random molecular structure allows it to behave in many ways like a semi-permeable super cooled liquid, with a melting point of about 1700 degrees C. Yet glass artifacts are not produced at anywhere near this temperature. The melting point of silica can be reduced with the addition of a glass modifier that will reduce the temperatures needed to melt silica to around 600 or 700 degrees C. They include the alkaline oxides of sodium (Na) from soda ash and potassium (K) from potash. Other ingredients added to historic glass include calcium (Ca) from limestone or magnesium (Mg) to promote stability in the glass structure. Finally a metallic oxide is added as a colorant. Copper oxides produce blue and green glass, iron produces dark green. Many of the metallic oxides produce color variants that seem rather unpredictable, as color is dependent on an ion's oxidation or reduction state. Some colors produced by these ions seem to cancel other colors, hence the creation of opaque and clear glass. Since most sand is naturally contaminated with iron oxides much historic glass was produced of the dark green variety (Pearson, 1987: 101; Cronyn, 1990: 128, 129).

Historic glass seems to come in two varieties, potash-lime-silica and soda-lime-silica. Until the 17th century nearly all the glass produced was of the potash-lime-silica variety. The potash was produced from bone ash. In this older glass flint was often used for the glass former instead of silica and the glass often contained a high lead content. Leaded crystal is of this variety of glass (Plenderleith, 1971: 343). The addition of lead to produce clear glass seems a natural outcome of using lead over glazing in pottery.

By the 19th century most of the glass produced was of the soda-lime-silica variety. The proportions of this glass varied from glass works to glass works, but ranges in the written sources from 60% to 73% silica, 15% to 5% lime, and 25% to 22% soda (Plenderleith, 1971: 344; Singley, 1988, 22).

Most conservators accept that glass deterioration, though not fully understood, is accelerated by the hydrolysis of glass modifiers and stabilizers. Glass structure consists of three-dimensional negatively charged poly-silicate ions. The silicate charge in the structure is offset by the positively charged modifiers (K^+ and Na^+) and stabilizers (Ca^+ and Mg^+). Should the positive ions Ca^+ or Mg^+ leach out of the structure, or not be present is sufficient quantity to begin with (as in badly made glass), the K^+ and or Na^+ will also begin to leach out. In the atmosphere these leaching ions first combine with water and become hydroxides. After drying they combine with carbon dioxide from the atmosphere to form sodium and potassium carbonates. As more moisture gathers on the glass the free positively charged hydrogen ions from the water migrate into the glass structure and take the place of the migrating modifiers and stabilizers thereby hydrating the glass. The potassium or sodium carbonates continue to absorb more moisture from the air and the process continues. This is called "glass disease" or "weeping or

Figure 37. Devitrifying glass bottle bottom. Photograph by Chriss Valvano.

sweating glass (Plenderleith, 1971: 344, 345)." Eventually the glass will take on a multihued iridescence and later begin to flake and devitrify.

Underwater the process is accelerated by the presence of large quantities of free hydrogen ions that will take the place of the migrating modifiers and stabilizers. Additional hydrogen ions produced by corroding metals or pollutants will further exacerbate glass hydration. When glass begins to devitrify it becomes susceptible to salt infiltration. Sodium chloride and water will penetrate the newly forming layers in the glass. If it is recovered and allowed to dry the salt will crystallize and rapidly exfoliate the outer layers.

STONE (LITHICS)—THEORY

The connotation of ancient people wearing animal skins and carrying stone tools is inescapable in archaeology, we somehow picture them in our imaginations as primitive, perhaps slow witted. The fact that we have survived as a species belies that fantasy. Nature honed the minds of our ever more clever and resourceful ancestors. Human use of lithic technology served us well for hundreds of thousands of years permitting humans to survive and prosper. Stone usage was sophisticated, each type of stone had a use, fully suited to its physical properties. Knapped projectile points are just one example of the art, beauty, and craftsmanship that went into early stone technology. When metals were discovered stone technology gradually shifts to architecture, building materials, concrete, roads, aqueducts, and a source of raw mineral resources for other industrial processes.

Geologists divide rock into three logical categories. These categories are extremely useful in archaeological conservation, for in a general sense they describe much concerning the physical qualities of the stone. Igneous rock includes those formed from volcanism both within and on the earth. Sedimentary rocks are layered, formed through gradual deposition of minerals in oceanic or fluvial environments. Metamorphic rock begins as igneous or sedimentary but is changed through heat and pressure in the earth.

Igneous Rock Examples

Granite	Impervious
Basalt	Impervious
Obsidian	Impervious
Gabbro	Impervious
Pumice	Porous

Archaeologically retrieved stone artifacts all fit into these categories and can range from projectile points and food preparation tools, to building materials and ballast stone. Conservation of these materials depends on their mineral composition, how they were formed, their porosity, hardness, strength, and thermal properties (Pearson, 1987: 103). Yet it can be seen, with some exceptions, that

Sedimentary Rock Examples

Breccia	Porous
Bituminous Coal	Porous
Flint	Porous
Limestone	Porous
Sandstone	Porous
Shale	Porous

Metamorphic Rock Examples

Anthracite Coal	Hard, not impervious
Dolomite	Hard, can be damaged in acid
Gneiss	Hard, not impervious
Quartzite	Hard, not impervious
Slate	Hard, not impervious
Marble	Hard, can be damaged in acid

igneous rock is similar in hardness and imperviousness to infiltration of salt and water to high-fired ware. Metamorphic rock is analogous to earthenware, while sedimentary rock has similar properties to low-fired ceramic wares.

As a general rule, though lithics are usually considered fairly stable materials, they degrade in the same manner as ceramic and glass. They can suffer from biological attack underground and underwater, they can be physically eroded by water and wind, and chemically if they are porous they can be infiltrated by salt, metallic salts, and sulfides. Rock, therefore, will be stabilized in similar fashion to ceramics and glass. And like ceramic and glass, the conservator/archaeologist's first and most important task is simply to identify the lithic type and theoretical decomposition from its description and properties.

RECOVERY AND STORAGE—METHODOLOGY

Dry recovered ceramics, glass, and stone should be safe for removal with cleaning and a rinse. It should be remembered, however, that porous artifacts have absorbed salts from the earth that may crystallize to the detriment of the artifact. A rinse on removal should obviate the surface salts and make the artifact safe for dry storage until the desalination rinse.

Siliceous materials recovered from underwater sites offer a greater challenge to the archaeologist/conservator. Like all materials recovered from the sea floor ceramics, glass, and lithics need to be kept wet. These materials are subject to calcarious concretion formation. If allowed to dry, these concretions will absorb carbon out of the atmosphere and harden, making it difficult to remove the concretion without damaging the artifact.

Additionally, all of the porous artifacts are subject to salt saturation and infiltration. On dehydration the salt can crystallize damaging the body and exfoliating any loose glaze or devitrified glass.

Salt saturation can also result in osmotic pressure differentials should the artifact immediately be immersed in fresh water. In order to lessen the risk that osmotic pressure differentials will exfoliate the outer layers of some ceramic, stone, and devitrified glass, it is recommended that the initial storage solution be 50% fresh water mixed with 50% salt water. After one week the solution can again be cut with a 50% fresh water infusion. After one more 50% fresh water change the salinity should be no more than 4.5 parts per thousand and the artifacts should be safe to store in 100% fresh water.

Material recovered from fresh water environments can remain stored in fresh water without worry concerning salt intrusion and osmotic pressure differentials. Some glass and stone, however, is adversely affected by acids and care should be taken that these materials are not stored for any length of time near corroding metal.

CONCRETION REMOVAL IMPERMEABLE ARTIFACTS—METHODOLOGY

Dry recovered artifacts will only be concreted if they were interred near iron artifacts and absorbed into the iron concretion. In this case the iron artifact should be treated first and the siliceous object removed from the concretion in the course of treatment. After removal, however, the ceramic, glass, or lithic can and should be examined and differentiated on the basis of its porosity. If it is a robust, non-porous artifact in the high fired, stable category it should follow the same treatment as that outlined below. If it is a porous, low fired, or devitrifying artifact it should be placed in the treatment outlined for concretion Removal—Porous Artifacts.

Acids can safely be used to remove concretion from high fired wares, most igneous, some metamorphic rock, and glass, provided it shows no signs of devitrification and is not left in the acid for extended periods (more than a week). Solutions of virtually any acid will remove the calcium carbonate concretions though the minimal intervention laboratory will be well supplied with citric acid. Soaks of 10% acid are effective in removing these concretions and can be mixed in the ratio of 10 grams of citric powder, to 90 ml of distilled water.

Long soaks in 10% sodium hexametaphosphate (Calgon) will also soften and help remove the concretion. However, if the concretion has been allowed to dry or is heavily contaminated with iron (cementite) the acid and Calgon soak will be only partially effective. Should this be the case only imaginative mechanical cleaning methods will prove productive. Dental tools and sharp scalpels will need to be employed in a vigorous manner. These robust artifacts should be up to the challenge without breakage.

CONCRETION REMOVAL POROUS ARTIFACTS—METHODOLOGY

Concretion removal for low-fired ceramics, devitrifying glass, some metamorphic rock, and sedimentary rock is a considerably more complex task than it was for the impervious materials in the previous section. In many cases with these more porous artifacts the concretion has penetrated under the glaze and into the body of the object.

A general acid soak is not recommended for porous siliceous materials. An acid soak will soften these materials and gas generation from carbon dioxide gas released from the dissolving calcium carbonate can damage delicate outer layers or loosen glazes from earthenware. In addition, the acid may dissolve iron from the glaze or body of a ceramic complicating or making impossible accurate radiographic spectral analysis (used to determine mineral content and thereby area of origin).

Fortunately lengthy soaks in sodium hexametaphosphate (Calgon) have proven gentle and effective in removing concretions from these more delicate objects. Experience has demonstrated that chelating agents like sodium hexametaphosphate will soften low-fired wares, so great care must be taken to support the artifact until it is dehydrated. Mechanical cleaning will be necessary with the Calgon soak to remove concretion, the Calgon will simply make it softer.

Stubborn concretions may require acid poultices produced by mixing acid and talcum powder together. This paste should be placed on the concretion rather than the artifact so as not to cause any general harm. Delicate mechanical cleaning will be needed to back up the poultice method.

STAIN REMOVAL IMPERMEABLE ARTIFACTS—METHODOLOGY

Metallic salt stains, rust red for iron and blue-green for copper, will most likely be removed from high fired ware, robust glass, and impermeable lithic during the acid soak concretion removal. This acid soak is the same treatment that would be prescribed to remove these stains if no concretion were present. As with the concretion removal, a 10% citric soak will suffice and will remove most stains within a week or two. Some tough stains residing on the roughened broken edge of a ceramic will need a citric paste poultice. This poultice is made with citric crystals, mixed with a few drops of distilled water. The paste is placed on the stain or vigorously scrubbed into the area with a toothbrush, results may take a few treatments.

Black organic stains produced by sulfides are removed by soaking the artifact in a strong 35% solution of hydrogen peroxide. Hydrogen peroxide of this strength should be handled with care as it can cause burns. Sulfide stain removal should take a few days to a week but some results will be viewed almost immediately.

STAIN REMOVAL POROUS ARTIFACTS—METHODOLOGY

Metallic salt stains in porous materials can be removed by encasing them in a paste of 10% citric acid mixed with talcum powder or citric acid soaked into a cotton bud poultice. It should be kept in mind that this artifact should not be allowed to dry while stain removal takes place. Salt crystallization is a possibility until the required desalination rinse is complete. Since the metallic stains will block the rinse it cannot be fully successful until the stains have been removed.

Chelating agents such as the disodium, trisodium, and tetrasodium salts of ethylenediaminetetraacetic acid (EDTA) have also been used successfully to remove stains from artifacts and are available from chemical supply stores. EDTA can be applied in the same manner as citric acid, in poultice form. Complexing agents such as EDTA or Calgon mobilize heavy metals from their insoluble compounds, so they can be rinsed from the artifact.[1] Again it should be noted that general soaks in chelating agents will soften low fired ceramic and must be done with great care.

Organic sulfide stains (black) are removed with a hydrogen peroxide soak. A 10% solution will keep gas evolution to a minimum. It may take several weeks for this weaker solution to remove the stain.

Devitrifying glass is a particular concern in this category. The sodium and potassium carbonates on the surface of the glass are hygroscopic and will continue to attract moisture from the atmosphere even after the glass has been dehydrated. In this instance a mild soap can be used to remove the carbonates before the artifact is given a final desalination rinse.

DESALINATION RINSE—METHODOLOGY

After concretion and stain removal all ceramics, glass, and lithic should undergo a lengthy desalination rinse in deionized or distilled water. This rinse will remove most of the remaining interstitial salt and is particularly indispensable to the more porous terra cotta, earthenware, sedimentary, and porous metamorphic rock whose bodies have absorbed a good deal of this harmful solute. Without the rinse any salt inside the artifact will have a tendency to rise to the surface and crystallize, flaking off surface detail and glaze.

Rinse tanks can be fabricated of any container that will hold the artifacts and a sufficient amount of distilled water to cover them. Experience has demonstrated, however, that stirring devices such as pumps or circulators of virtually any kind,

[1] A Chelating Agent is an electron donor compound with more than one atom that may be bonded to a central metal ion at one time to form a ring structure. Chelates of EDTA greatly increase the migration of heavy metal ions from existing compounds. These agents are not specific for a single metal ion. When EDTA complexes with a metal ion from the oxide layer, the resultant complex is more water soluble (Personal communication with M. Kuehl, chemistry consultant).

will speed the rinse process by a large factor. Stirring devices keep pockets of high concentration salt from building in the interstitial artifact—rinse water boundary. The high salt concentration in these small crevices can block the release of more salt and slow the rinse time.

Rinse times are variable depending on the type of rinse tank used. Friable and delicate objects should undergo static rinse in an un-circulated container. Static immersion rinses of two months with weekly water changes for porous artifacts is not out of the question. Agitation tanks require less time, with a month of rinse time and weekly water changes recommended for badly contaminated porous objects.

Fresh water recovered artifacts have not been subject to salt infusion and, therefore, require no fresh water rinsing.

CONSOLIDATION—METHODOLOGY

Only those delicate and friable artifacts that are actually loosing surface detail should be considered for consolidation. These artifacts should be dehydrated in a solvent solution to lessen the surface tension effects of drying water as it pulls on delicate surface structure. An hour long soak in used alcohol, acetone, or toluene will suffice for the initial dehydration followed by one more hour-long soak in fresh 100% solvent. Solvent baths should be covered at all times and located far from any flammable ignition source.

Consolidation or the physical stabilization of the surface of a delicate artifact can then be accomplished by soaking the dehydrated artifact in a mixture of Paraloid B-72 (glue) in toluene, or PVA (polyvinyl acetate) in any solvent. The percentage mixture of these consolidating substances in solvent depends on the amount of penetration needed. The more penetration the more solvent, the more surface consolidation the less solvent. Mixtures as high as 50% solvent and 50% glue can be used. A few hours soaking in the consolidant mixture should suffice for the glue and solvent to penetrate the surface and into the body of the object. The artifact can then be carefully removed and placed in a vapor chamber. In this instance a vapor chamber can be any sealed plastic container that is not much larger than the object that is being treated. Three days of slow drying is sufficient time to prevent the consolidant from wicking to the surface of the object.

It should be noted that the consolidants listed here are not completely impervious to moisture penetration. It is possible that devitrifying glass can still absorb water from the atmosphere and continue to decompose.

Devitrifying glass is a particular problem in conservation. Washing the sodium and potassium carbonates from the surface (in Stain Removal Porous Artifacts) will slow the ion migration from the glass, but will not stop it. New treatments like silicon oil (Smith, 2003) show promise and short term good results by blocking the ion exchange and consolidating the surface. But silicon oil treatment is irreversible and has not proven itself over long duration. Until a treatment arrives that satisfies the conservation codes stated in chapter 1, the best treatment for devitrifying glass is to store it in a very low relative humidity, even if it has been consolidated with B-72 or PVA. This is easily accomplished by placing the treated glass in a small chamber or container supplied with a desiccant (moisture absorbing salt).

DEHYDRATION—METHODOLOGY

Final dehydration is accomplished by removing the artifacts from the desalination rinse and allowing them to air dry. As mentioned in the consolidation section, particularly delicate artifacts can be solvent dried in used alcohol for a one-hour soak followed by another similarly timed soak in 100% solvent. This solvent dehydration will lessen the surface tension effect of water and allow the object to air dry with little or no surface damage.

CONCLUSION

Misidentification of ceramics and pottery is always possible so long as stains hide the true color and nature of the piece and tool marks and mold marks are obscured by concretion and staining. Fortunately conservation remedies for these problems are not difficult and will aid in the micro-excavation and evaluation of siliceous artifacts.

Yet ceramic, glass and lithic artifacts pose unique problems for the archaeologist/conservator. In many ways the robust nature of these objects work against them, in that they deceive the archaeologist/conservator, or curator into thinking that the materials are indestructible and need no stabilization. This is simply not the case, many of these artifacts have been damaged over time in a great many ways and many will continue to degrade in the laboratory or museum if the artifact is not treated and stabilized for storage.

CHAPTER 6: ARCHAEOLOGICAL CERAMIC, GLASS, AND STONE

"An Act of Faith and the Restorer's Art." *Smithsonian* Vol. 30 No. 8, 1999, pp. 76–85. [ECU788]

Andre, Jean-Michel. *The Restorer's Handbook of Ceramics and Glass*. Toronto: Van Nostrand Reinhold, 1976.

Beaubein, L.A. "Ceramics, Paper, Fabrics, Photographic Materials, and Magnetic Tape." *Seawater Corrosion Handbook*. M. Schumacher, ed. Park Ridge, New Jersey: Noyes Data Corporation, 1979: 466–472.

Bhargav, J.S., R.C. Mishra, and C.R. Das. "Environmental Deterioration of Stone Monuments of Bhubaneswar, the Temple City of India." *Studies in Conservation* Vol. 44 No. 1, 1999, pp. 1–11. [ECU-710]

Biser, Benjamin. *Elements of Glass and Glassmaking*. Pittsburgh: Glass and Pottery Publishing, 1899.

Brill, Robert H. "Ancient Glass." *Scientific American* (November 1969): 120–130. [ECU-269]

—. "Analyses of Some Finds from the Gnalic Wreck." *Journal of Glass Studies* 15 (1973): 93–97. [ECU-131]

—. "Crizzling—A Problem in Glass Conservation." *Conservation in Archaeology and Fine Arts*. (International Institute for Conservation, London) (1975): 121–134.

—. "The Use of Equilibrated Silica Gel for the Protection of Glass with Incipient Crizzling." *Journal of Glass Studies* 9 (1978): 1001–1118.

Brus, Jiöi and Petr Kotlik. "Consolidation of Stone by Mixtures of Alkoxysilane and Acrylic Polymer." *Studies in Conservation* Vol. 41 No. 2, 1996, pp. 109–119. [ECU-711]

Buys, Susan and Victoria Oakley. *Conservation and Restoration of Ceramics*. London: Elsevier, 1996.

Caldararo, Niccolo. "Conservation Treatments of Paintings on Ceramic and Glass: Two Case Studies." *Studies in Conservation* Vol. 42 No. 3, 1997, pp. 57–164. [ECU-712]

CCI Laboratory Staff. "Care of Black-and-White Photographic Glass Plate Negatives." *CCI Notes* (Canadian Conservation Institute) 16/2 (1988). [ECU-442]

CCI Laboratory Staff. "Care of Ceramics and Glass." *CCI Notes* (Canadian Conservation Institute) 5/1 (1991). [ECU-352]

CCI Laboratory Staff. "Care of Argillite." *CCI Notes* (Canadian Conservation Institute) 12/1 (1992). [ECU-461]

Clow, A. & N.L. Clow. "Ceramics from the 15th Century to the Rise of the Staffordshire Potteries." *A History of Technology Vol IV: The Industrial Revolution c. 1750–1850*. Oxford University Press, London 1958. [ECU-565]

Collins, Chris. *Care and Conservation of Palaeontological Material*. London: Butterworths, 1995.

Cronyn, J.M. *The Elements of Archaeological Conservation*. London: Rutledge, 1990.

Crossley, D.W. and F.A. Aberg. "Sixteenth Century Glass-Making in Yorkshire: Excavations at Furnaces at Hutton and Rosedale, North Riding, 1968–71." *Post Medieval Archaeology* 6 (1972): 107–59.

Coysh, A.W. & Henrywood, R.K. *The Dictionary of Blue and White Printed Pottery, 1780–1880*. Vol. 1. Woodridge, Suffolk: Antiques Collectors Club Ltd. 1982.

Davison, Sandra. *The Conservation of Glass*. London: Butterworths, 1982.

Dei, Luigi, Andreas Ahle, Piero Baglioni, Daniela Dini, and Enzo Ferroni. "Green Degradation Producst of Azurite in Wall Paintings: Identification and Conservation Treatment." *Studies in Conservation* Vol. 43 No. 2, 1998, pp. 80–88. [ECU-713]

Dimes, F.G. and Ashurst, J. Conservation of Building and Decorative Stone. London: Elsevier, 1998.

Frank, Susan. *Glass and Archaeology*. London: Academic Press, 1982.

Gilles, William B. "Bottles from the Sea, Part I.: Recovery and Treatment." *Skin Diver* (April 1986): 114–119. [ECU-428]

Glass Glossary. Parks Canada—National Historic Parks and Sites, 1989. [ECU-433]

Godden, Geoffrey A. New handbook of British Pottery & Porcelain Marks. London: Barrie & Jenkins, 1968.

Goodenough, Robert D. "Method and Composition for Removing Iron Stains from Porcelain." (U.S. Patent Number 3,721,629): 1–4.

Guldbeck, Per E. "Ceramics: Care and Conservation." *The Care of Antiques and Historical Collections*. Nashville: AASLH Press, 1986. [ECU-321]

—. "Glass: Problems and Solutions." *The Care of Antiques and Historical Collections*. Nashville: AASLH Press, 1986. [ECU-322]

Hamilton, Donny. *Methods for Conserving Underwater Archaeological Material Culture*. College Station, TX: Texas A & M University, 1998.

Hawley, J.K., Kawai, E.A., and C. Sergeant. "The Removal of Rust Stains from Arctic Tin Can Labels Using Sodium Hydrosulfite," *Journal of the IIC-CG*, 6 (1 & 2) (1981): 17–24.

Hodges, Henry. *Artifacts: An Introduction to Early Materials and Technology: Chapter 2: Glazes*. John Baker, London 1964. [ECU-566]

Howie, M.P. Frank. *Care and Conservation of Geological Material, Minerals, Rocks, Meteorites and Lunar Finds*. London: Elsevier, 1992.

Hunter, Robert, ed. *Ceramics in America*. London: Chipstone Foundation, 2001.

Ilik-Yürüksoy and Olgun Güven. "The Preservation of Denizlilimestones by in-situ Polymerization." *Studies in Conservation* Vol. 42 No. 1, 1997, pp. 55–60. [ECU-714]

Jedrzejewska, Hanna. "Removal of Soluable Salts from Stone." *Proceedings* (IIC Conference on the Conservation of Stone & Wooden Objects, New York) 1 (1971): 19–33.

—. "The Kern Effigy: Its Discovery, Interpretation, and Infield Preservation." *Curator* Vol. 36 No. 1, 1993, pp. 66–78. [ECU-715]

Koob, S.P. "The Removal of Aged Shellac Adhesive from Ceramics." *Studies in Conservation* 24 (1979): 134–35. [ECU-181]

Koob, Stephan. "Obsolete Fill Materials Found on Ceramics." *Journal of the American Institute for Conservation* Vol. 31 No. 2, 1992. [ECU-716]

Koob, Stephan and Won Yee Ng. "The Desalination of Ceramics Using a Semi-Automated Continuous Washing Station." *Studies in Conservation* Vol. 45 No. 4, 2000, pp. 265–273. [ECU-717]

—. "The Use of Paraloid B-72 as an Adhesive: Its Application for Archaeological Ceramics and other Material." *Studies in Conservation* 31 (1986): 7–14. [ECU-306]

Kovel, Ralph and Terry Kovel. "Bottles." *Know Your Antiques*. New York: Crown Publishers, 1967: 100–115. [ECU-0367]

MacLeod, Ian D. "Desalination of Glass, Stone, and Ceramics Recovered from Shipwreck Sites." *ICOM Proceedings* (Sydney) (1987): 1005–1007.

Mack, Robert C. "Brick and Stone Preservation: The First Steps." *Early American Life* (June 1977): 52–57. [ECU-424]

Matero, Frank G. and Albert Tagle. "Cleaning, Iron Stain Removal, and Surface Repair of Architectural Marble and Crystalline Limestone: The Metropolitan Club." *Journal of the American Institute for Conservation* Vol. 34 No. 1, 1995. [ECU-718]

McKearin, George S. and Helen McKearin. *American Glass*. New York: Crown Publishers, 1941.

Mibach, Lisa. "The Restoration of Coarse Archaeological Ceramics." *Proceedings* N. Brommellie and P. Smith, eds., (1975 Stockholm Congress, International Institute for Conservation of Historic and Artistic Works) (1975): 63–68.

Moyer, Cynthia and Gordon Hanlon. "Conservation of the Darnault Mirror: An Acrylic Emulsion Compensation System." *Journal of the American Institute for Conservation* Vol. 35 No. 3, 1996. [ECU-719]

Munnikendam, R.A. "Preliminary Notes on the Consolidation of Porous Building Materials by Impregnation with Monomers." *Studies in Conservation* 12 (1967): 158–62. [ECU-345]

Newton, Roy and Sandra Davison. *The Conservation of Glass*. London: Butterworths, 1989.

Nylander, Carl. "S.O.S. for Ancient Monuments." *Archaeology* 41 (1988). [ECU-348]

Oddy, W.A. and H. Lane. "The Conservation of Waterlogged Shale." *Studies in Conservation* 21 (1976): 63–66.

Olive, J. and C. Pearson. "Conservation of Ceramics from Marine Archaeological Sources." *Conservation in Archaeology and the Applied Arts* (Proceedings of the 1975 Stockholm Congress, International Institute for Conservation, London) (1975): 63–68.

Pannel, Jane. "Conservation of Glass in Bodrum Museum of Underwater Archaeology." *Uluslararasi Anadolu Cam Sanati Sempozumu Kitabi* (1990): 47–50, 113–114.

Pauter, Ian. "An Investigation into the Drying and Consolidation of Wet Glass Recovered From the *Mary Rose*." *ICOM Proceedings* (Sydney) (1987): 1013–1016.

Pearson, Colin. "Deterioration of Ceramics, Glass, and Stone." *Conservation of Marine Archaeological Objects*. Butterworths, London, 1987. [ECU-562]

Pessoa, J. Costa, J.L. Farinha Antunes, M.O. Figueiredo, and M.A. Fortes. "Removal and Analysis of Soluable Salts from Ancient Tiles." *Studies in Conservation* Vol. 41 No. 3, 1996, pp. 153–160. [ECU-720]

Plenderleith, H.J. & Werner, A.E.A. The Conservation of Antiquities and Works of Art. London: Oxford University Press, 1971.

"Preservation of Historic Masonry." *Journal of the Association of Preservation Technology* Vol. 26 No. 4, 1995. [ECU-721]

Rodgers, Bradley A. *The East Carolina University Conservator's Cookbook: A Methodological Approach to the Conservation of Water Soaked Artifacts.* Herbert R. Paschal Memorial Fund Publication, East Carolina University, Program in Maritime History and Underwater Research, 1992. [ECU-0402]

Saleh, Saleh A., Fatma M. Helmi, Monir M. Kamal, and Abdel-Fattah E. El-Banna. "Study and Consolidation of Sandstone." *Studies in Conservation* Vol. 37 No. 2, 1992, pp. 93–104. [ECU-722]

Sease, Catherine. *A Conservation Manual for the Field Archaeologist.* (Archaeological Research Tools Vol. 4, Institute of Archaeology, University of California, Los Angeles) (1987). [ECU-221]

Singley, Katherine. *The Conservation of Archaeological Artifacts From Freshwater Environments.* South Haven, MI.: Lake Michigan Maritime Museum, 1988.

Smith, Wayne C. *Archaeological Conservation Using Polymers, Practical Applications for Organic Artifact Stabilization.* College Station, Texas A & M University Press, 2003.

Staniforth, Mark & Nash, Mike. *Chinese Export Porcelain from the Wreck of the Sydney Cove (1797).* Tasmania: The Australian Institute for Maritime Archaelogy, Inc., 1998.

Stein, Rena E., Jocelyn Kimmel, Michele Marincola, and Friederike Klemm. "Observations on Cyclododecane as a Temporary Consolidant for Stone." *Journal of the American Institute for Conservation* Vol. 39 No. 3, 2000. [ECU-723]

Sullivan, Catherine and Olive Jones. *The Parks Canada Glass Glossary* (Canada Parks Service, Ottawa) (1989).

Thomson, Garry, ed. *Conservation of Stone. Proceedings* (IIC Conference on the Conservation of Stone and Wooden Objects, New York) 1 (1971).

Warren, John. *Conservation of Brick.* London: Elsevier, 1998.

Weier, Lucy E. "The Deterioration of Inorganic Materials Under the Sea." *Archaeological Bulletin* (University of London, II) (1974): 131–63.

Wenker, Bert and Ellen Wenker. *The Main Street Pocket Guide to North American Pottery and Porcelain.* Pittstown, New Jersey: Main Street Press, 1985.

Williams, Nigel. *Porcelain—Repair and Restoration.* London: British Museum, 1983.

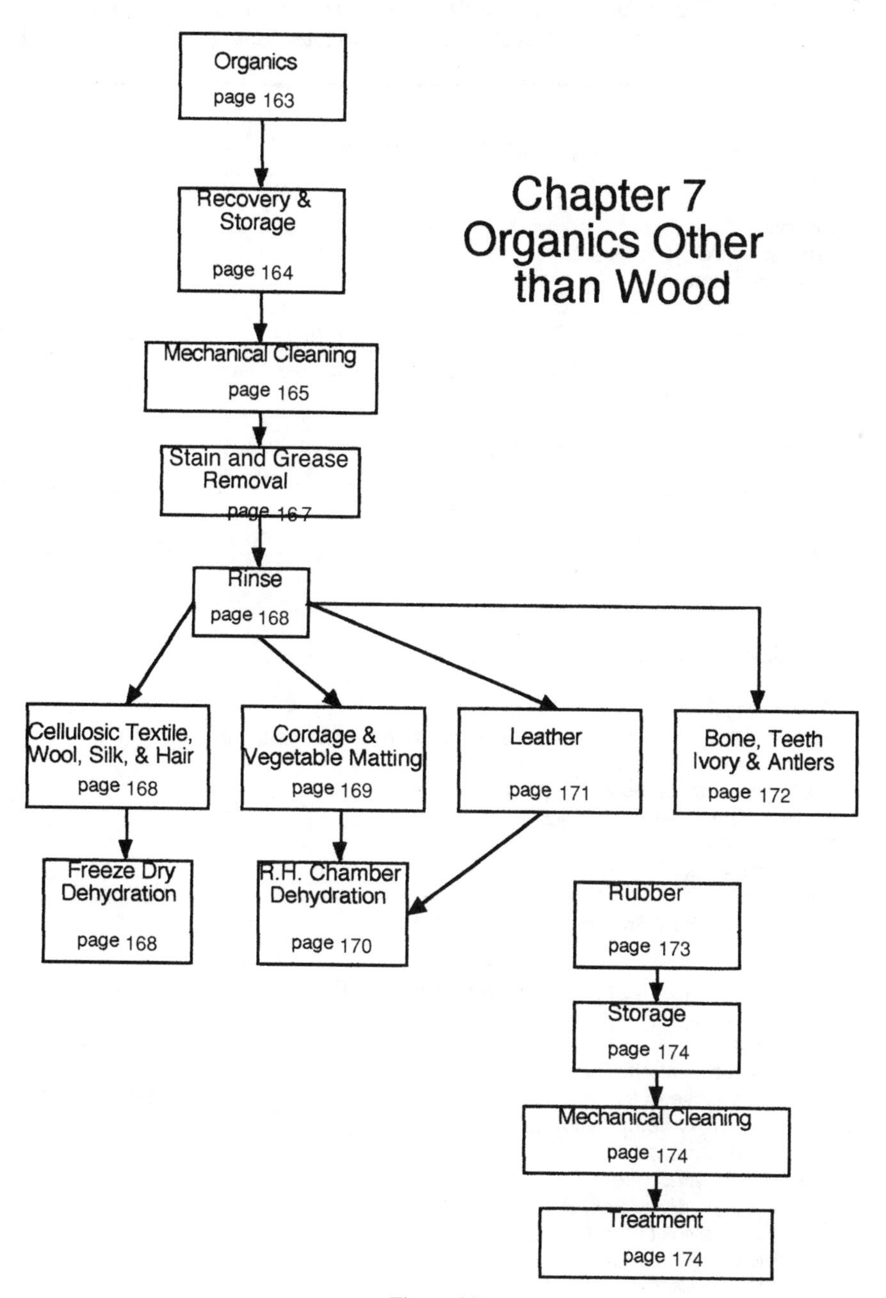

Organics
page 163
Recovery & Storage
page 164
Chapter 7
Organics Other than Wood
Mechanical Cleaning
page 165
Stain and Grease Removal
page 167
Rinse
page 168
Cellulosic Textile, Wool, Silk, & Hair
page 168
Cordage & Vegetable Matting
page 169
Leather
page 171
Bone, Teeth Ivory & Antlers
page 172
Freeze Dry Dehydration
page 168
R.H. Chamber Dehydration
page 170
Rubber
page 173
Storage
page 174
Mechanical Cleaning
page 174
Treatment
page 174

Figure 38.

Treatments for Organics Other than Wood
Represented in Conservation Literature

Stain Removal

Acetic acid
Amonia
Amoniated Peroxide Bleach
Amonium Citrate
Amonium Oxalate
Chloramine - T
Citric acid 167
EDTA (di, tri, & tetrasodium salt) 167
Formic acid
Hydrochloric acid
Hydrofluoric acid 167
Hydrogen Peroxide 167
Hydrogen Peroxide + Amonia
Morpholine
Non-Ionic detergent 168
Oxalic Acid 167
Phosphoric acid
Potassium Permanganate
Pyridine
Sodium Bisulfite
Sodium Citrate + Sodium Dithionite + Sodium hydrogen carbonate
Sodium Dithionite
Sodium Hexametaphosphate (Calgon)
Sodium Hydrosulfite + Amonia
Sodium Sulfite
Sulphurous acid
Tartaric acid
Thioglycollic acid + Sodium Dithionite + DTPA

Consolidation

Cellulosics
- Carboxymethyl cellulose
- Ethulose 400 169
- Hydroxymethyl cellulose

Polyvinyl Alcohol
PVA (Polyvinyl acetate)
- Acryloid/ Paraloid B-72 171
- Bakelite AYAF & AYAT

PVB (Plolyvinyl Buterol)
Polyvinyl Pyrrolidone
Rhoplex
Silicon Oil
Soluble Nylon
White Glue (PVA) 173

Figure 39.

Leather Treatments

Bavon ASAK ABP (Solvent)
Bavon 5205 (Water)
British Museum Leather Dressing
Castor Oil
Deutches Leder Museum Emulsin
Dinolene 1230 - B
Dymsol D
Dymsol S
Glycerol
Guildhall Museum Dressing
Lexol...
Neatsfoot Oil
Neutralfatt SS
Petrolatum
Polyethylene Glycol (PEG)
Silicon Oil
Vegetable Oil

Dehydration Treatments

Atmospheric Freeze Drying.. 168, 170, 171

Slow Drying (Humidity Chamber).. 170, 171

Solvent Drying ....removes fats & oils

Vacuum Freeze Drying...not cost effective

Pearson, 1987: Chapt. 8; Plenderleith, 1971: 110-111; Singley, 1988: 70-89; Hamilton, 1999: Files 3, 7,. 8)

Figure 39. (continued)

Chapter 7

Organics Other than Wood

Organics other than wood is an immense category of materials that include all artifacts produced from plant fiber, plant byproducts, plus animal tissue and bone. Although each of these materials deserves a chapter unto itself (if not a book), there are enough similarities of treatment within these materials to group them together, into one chapter. As well as being an extremely large group of materials, organics are extremely complex. In this light, a discussion of animal DNA protein synthesis within the ribosomes, and carbohydrate formation from dark and light cycle photosynthesis in plants, would only cloud conservation issues and make the text needlessly complex.[1] Therefore, the description of biological structure in organics will be limited to those salient features and structure that actually affect the outcome and choice of conservation treatment.

Surprisingly, many organics other than wood survive the centuries quite well. This is surprising in that the earth's ecosystems are generally quite adept at breaking down organic material into raw nutrients and minerals. The landscape and ocean floor would be quite different and very cluttered if nothing decayed. Yet internment at archaeological sites seems to offer an alternative to the general breakdown rules. Leather and bone rank among the most durable archaeological materials if interred under the right circumstances, while seemingly delicate items such as textiles and cordage survive inundation and benthic burial far better and more often than archaeologists and conservators would normally have believed.

All of these materials fit under the heading of carbon based organic fibrous polymers. Simply put, they are materials created by living cells made of strings of molecules, protein from animals and cellulose or other polysaccharides from plants. Cellulose is made of glucose (a type of sugar) monomers chained together to form a polymer or polysaccharide. Animal protein can be divided into two basic types, collagen and Keratin. Collagen, such as skin, leather, bone and antlers consist of protein deposits outside the animal cell. Keratin forms such material as hair, horns, shell, or wool. Keratin is an interior cell deposit.

Cellulosic material on the other hand, can often be divided into soft fibers such as flax, cotton, baste, linen, ramie, sisal, jute, and kapok and hard fibers such

[1] Should the conservator be interested, some of the better discussions of organic formation include (Pearson, 1987: Chapters 2 & 8; Cronyn, 1990: Chapter 6; Plenderleith, 1971: Part 1; Hamilton, 1999: files 3, 7, 8).

as manila, hemp or caroa. Other important cellulosic materials found on archaeological sites may include paper, straw, corn, cornhusks, as well as other grains, nuts, and fruit pits, and occasionally pine needles used for utilitarian items such as baskets and brooms. Plant products such as pitch, tar, and other resins plus rubber are not cellulosic materials but polymeric products of the living process in trees.

Left on the surface of the ground, most of these organic materials will quickly decay or be consumed by a variety of animals and bacteria. Bones and antlers left in wooded environments are of particular interest to other animals for their calcium salts. Yet most organic materials can survive quite lengthy burials, in the ground or the sea bottom. Often these materials reach a state of equilibrium with the site in which they are interred. Though most will continue to slowly decompose, anaerobic decomposition is much less efficient than aerobic respiration used by surface level bacteria and air breathing animals. Oxygen, pH, and Eh, levels in the ground or sea bottom complicate decomposition as do the presence of tannins or metallic salts. Decomposition also depends on the amount of interstitial water present at a site. Desert and dry sites can exhibit excellent preservation of organic material while wet sites demonstrate preferential preservation. In all, therefore, the only factor that is nearly as complex as a study of organic material, is site formation process of their burial.

All in all, therefore, it is more efficient to tackle organic materials by their similarities of treatment, rather than go into the formation and breakdown of each type of material as has been done for wood, metal, and siliceous artifacts. This methodological approach may fall short of explaining why a particular method is chosen in each case, but will prove reasonably efficient.

STORAGE—METHODOLOGY

It should be remembered that removing any artifact from the nearly static equilibrium of its internment site changes its' decomposition dynamics. Most artifacts, particularly organic materials will begin rapid decomposition after excavation and must be considered to be under micro-biotic attack. It is the job of the archaeologist/ conservator to see to it that the artifact and the data that it represents arrive safely at the laboratory and begin stabilization. As with all other materials, stabilization of organic materials begins with its recovery and storage.

Dry recovered organic artifacts are often excavated along with the surrounding soil in a block recovery. In this way, the organic artifact, textile, leather, bone, will remain supported by the surrounding soil until micro-excavation can carefully release the item at the laboratory. Block recoveries are easily boxed for transportation provided the block fits satisfactorily into the transportation box. All recovered organic artifacts should be considered delicate. If recovered individually, they should be stored in plastic bags and supported within a transportation box in as close to the same humidity as their natural environment as possible. If the artifacts do not fit snuggly within their transport containers they should be packed so that they don't slide around. Quite often a great deal of damage can be visited on any artifact that is free to move inside a transport box.

In a like manner wet recovered or sea bed recovered organics should be kept damp, but unlike wood, they need not be left in containers of free standing water. Most organic artifacts have had their cellular structure altered during processing or manufacture and can no longer be affected by cellular collapse or osmotic pressure differentials. Therefore, they are best stored in only enough fresh water to keep them damp. The agitation created by transporting freestanding water in a storage container can turn organic materials to pulp. This is particularly true of cordage. Over the centuries it can be degraded to the consistency of soft paste.

The best laboratory storage for organic materials would recreate the anaerobic, dark, and temperature moderated conditions that exist in the soil or sea floor, yet recreating a non-oxygenated storage facility may be difficult and impractical. However, airtight plastic storage containers will work nicely on a case by case basis. These can be flooded with nitrogen if the artifact is valuable enough to warrant this small, added expense. Nitrogen makes up 78 percent of the earth's atmosphere and is relatively inert and non-toxic. A bottle of nitrogen opened directly over and within inches of an open box will spill into it displacing the oxygen rich atmosphere. The lid is placed on the artifact, and the box is for all intents and purposes oxygen free. Nitrogen can be purchased at any home brewing or bottling store.

In addition to air tight, possibly oxygen free storage, organic materials should be refrigerated, and depending on the material, they may be kept frozen. Textiles, wool, silk, and hair plus vegetable matting are all safe to freeze. Cool temperatures retard the metabolic rate of the micro-biota that will attempt to break down the organic material. Though refrigeration and freezing will not stop decay entirely, it will be greatly slowed. Refrigerators are fairly inexpensive and certainly trouble free. All laboratories that deal with organic artifacts should be supplied with a refrigerator/freezer.

Finally, custom designed supports at this stage of the treatment will help mitigate the physical damage incurred by simply transporting an artifact through its different treatment procedures. These supports can include nylon mesh artifact stockings placed around an artifact to surround and support it without interfering with treatment, or various imaginative combinations of plexi-glass and wood. Once placed in its support, an artifact should not have to be rearranged. In many cases, therefore, the artifact has to be shaped as the conservator would like it to look when it is conserved.

MECHANICAL CLEANING—METHODOLOGY

Often the silt that has helped save an organic artifact adheres to it quite strongly. Unwanted silt, clay, and sand can usually be rinsed and cleaned using mechanical cleaning procedures and a great deal of patience. At the lab the artifacts are carefully released from their embedded soil almost one particle at a time.

Micro-excavation and cleaning is necessarily a mechanical process in which the conservator/archaeologist utilizes scalpels, dental picks and other tools for the delicate work. Imagination can also be helpful, fabricated tools can make the delicate job of disinterring the artifact easier. Below is an example of a table that can be flooded to a depth of one inch and is back-lit. Back-lighting and a 3x lighted hobby lens allow for good scrutiny of the material while the cleaning process takes place. Soil and dirt are removed on this table with an I.V. needle attached to a tiny variable speed water pump. As the dirt and sand is pried off with the needle, the water stream washes it out of the way. On the other side of the table is a vacuum hose of the same small gauge tubing that supplies the I.V. needle with its water stream. The vacuum hose can be used to remove and displace pieces of dirt or move the fabric without touching it. I.V. needles make an excellent micro-excavation tool as they are sharp and well shaped for prying particles and dirt loose.

It is imperative during cleaning that textiles and other flat or woven artifacts and fabrics be unfolded and fully spread on a supporting structure. This is not easily accomplished until each layer of dirt is removed and a layer of fabric is unfolded. Unfolding fabric is more easily accomplished on a table that can be flooded. Water buoyancy helps support the weight of the fabric while it lubricates the process and

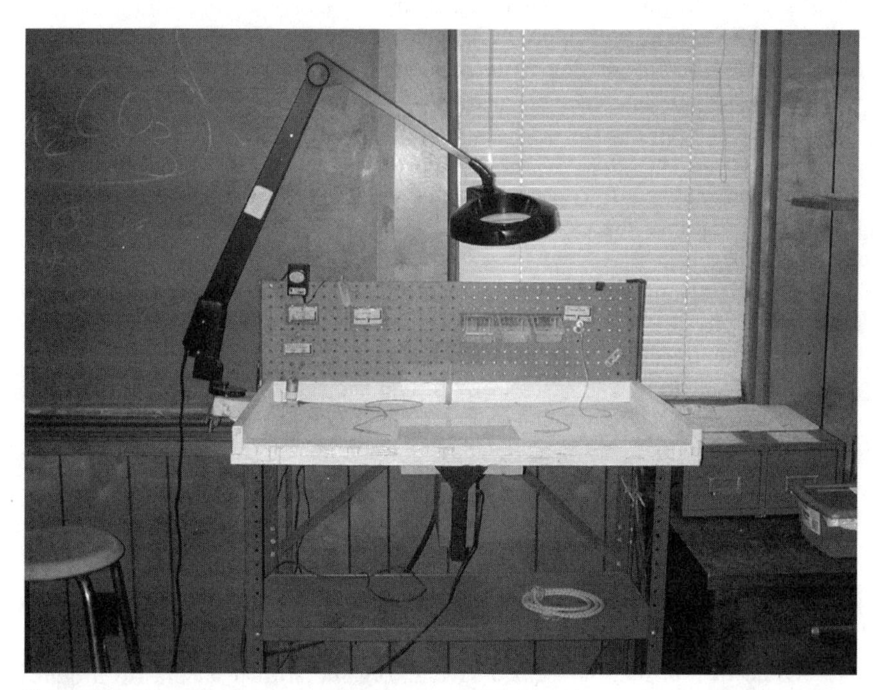

Figure 40. This textile and cordage cleaning table can be flooded with water to a depth of one inch Cleaning and unfolding textiles is accomplished with the aid of a small pump (on left) that produces a water stream. The other flexible tubing produces a vacuum. Photograph by Chris Valvano.

prevents breakage at the fold line. Between and after each cleaning session the artifact should be returned on its support to the refrigerator.

Mechanical cleaning is probably the most important stage in the micro-excavation of an organic artifact. It is here that the true nature and form of a textile or fabric can be realized. What was once a formless mass may suddenly take a shape and become a recognizable pattern. Other objects may also be recognized only after the cover of dirt and soil has been removed.

STAIN AND GREASE REMOVAL—METHODOLOGY

Most organic artifacts may exhibit two basic types of stain. Black organic stains from sulfides are quite damaging, as are metallic salt stains from nearby corroding metals. Iron stains will be rust orange while copper stains will be blue or green. Stains and the compounds that make them up are invariably damaging to organic material and should, if possible, be removed.

Sulfide stains are removed from most organic fabrics and polymers with a 3% to 10% hydrogen peroxide soak. Great care should be taken during this chemical wash to inspect the artifact at short intervals. Hydrogen peroxide H_2O_2 is a bleaching agent and can be very harmful to some textiles, actually dissolving soft cellulosic fibers in a relatively short time of several hours. The archaeologist/ conservator may have to decide the ratio of probable good to possible harm for any given artifact. All treatments and chemical washes should be attempted on small areas before the entire object is cleaned. Hydrogen peroxide soaks should be done with observers present and last no more than several hours per wash, per day. The artifact should be rinsed in distilled water after each soak and returned to the refrigerator afterward.

Metallic corrosion stains are removed in dilute acid solution or a chelating agent such as the disodium, trisodium, or tetrasodium salt of EDTA. Citric, oxalic, or hydrofluoric acids can be used in 3% to 10% concentrations (hydrofluoric acid does have some carcinogenic health concerns and should never be used without gloves and good ventilation). If an artifact is made of a material that is vulnerable to acid soaks (such as bone, teeth, ivory, and antler) a 10% trisodium or tetrasodium salt EDTA solution offers a neutral or basic pH alternative. As with hydrogen peroxide, great care must be taken to insure the optimal beneficial treatment time by subjecting a small portion of an object to the treatment, before submerging the entire artifact in the chemical wash. As always the possible benefits must outweigh the risks before a procedure continues.

Acid or EDTA treatments may last for several hours per treatment, per day. The stain removal should continue until the stain is gone or until the conservator believes the repeated soaks are damaging the artifact. Total treatment time may be measured in weeks. Sulfide and metallic stain intrusion is insidious and may take years to damage an artifact, but the damage will take place nonetheless. As with the hydrogen peroxide soak, the artifact should be rinsed in distilled water after each treatment and returned to the refrigerator between treatments.

When the acid wash is complete and the stains have been removed, a one hour buffering rinse in mild 1% or 2% sodium bicarbonate solution will neutralize any acid that may have soaked into the artifact. If grease is present in the material it can be removed with a 5% solution of non-ionic detergent (available through chemical suppliers). Buffering or grease removal should be followed by a distilled water rinse and all artifacts that have gone through this process, including dry recovered artifacts should remain stored in distilled water and placed in the refrigerator until the dehydration process. This will prevent damage from repeated drying.

RINSE—METHODOLOGY

After stain removal is complete a final static rinse in distilled water should last for several hours and the water should be changed at least once depending on the relative surface area of an artifact. The courser the texture of an object, the greater the surface area and the greater the need to change the rinse water. An overnight rinse is not out of the question. No agitation is necessary for this rinse as water movement may damage the artifact.

This rinse should remove all treatment and buffering solutions plus any salts that have penetrated the artifact. Handling the artifact should be kept to a minimal, by now it is only moved on the support constructed for it in the storage stage.

TREATMENT FOR CELLULOSIC TEXTILES, WOOL, SILK & HAIR—METHODOLOGY

Cellulosic textiles include any fabric or material fashioned of cellulose. This includes paper and woven materials comprised of cotton, kapok, baste, linen, flax, hemp, ramie, sisal, or jute. Though unrelated to cellulosic textiles except for their treatment, wool, silk (protein fibroin), and hair (protein keratin) are included in this category.

The treatment for these materials is very simple. They should be arranged in the fashion that the conservator believes' is best for display, and placed on their supporting structures in an ordinary freezer.

DEHYDRATION—METHODOLOGY

When stain removal and cleaning are completed cellulosic textiles plus wool, hair, and silk are ready for non-vacuum freeze-drying. Non vacuum freeze-drying is a simple yet effective method for the conservation of these materials. The textile is simply arranged on its support as it will be displayed, and placed in the freezer.

Within six weeks the water will have sublimated from the fabric leaving the dehydrated textile spread on its support. Sublimation occurs when a solid becomes a gas without going through a liquid phase. The liquid phase, with its increased capillary tension pressures, can be harmful to the artifact as it dehydrates.

Should it be deemed necessary, a 10% solution of glycerol in distilled water can be misted on the cellulosic fabrics, to keep them from becoming too dry and brittle. Glycerol will act as a humectant, helping add flexibility to the fibers. In a like manner, a dilute solution of lanolin can be atomized over woolen artifacts during freeze-drying. Lanolin is a natural wool fat derivative and will enter the wool and replace some of the fats lost through decay and hydrolysis.

After dehydration the artifacts are safe to store in a climatically controlled (air conditioned) building at 50% relative humidity. Containers and supports should be non-reactive and acid free (virtually any plastic). Cardboard boxes will release acid into the micro-atmosphere inside the box. Any artifacts, stored in cardboard boxes should be sealed in a protective plastic cases or bags and periodically checked to see that the humidity level in the bag or box does not promote mold.

TREATMENT FOR CORDAGE AND VEGETABLE MATTING—METHODOLOGY

Cordage includes all types of cellulosic rope, line, cannon shot wadding, and historic caulking. Vegetable matting includes anything from pine needle and grass woven baskets, to brooms or other utilitarian objects of like material.[2] Though these objects seem to be very delicate and unlikely to survive in the archaeological record, many have been recovered. However, biotic attack and hydrolysis usually leave cordage and vegetable matting recovered from both land and underwater sites in need of consolidation and structural support.

Most of these artifacts will be delicate and require physical supports and nylon mesh wraps (advocated in the Recovery and Storage Section). The nylon mesh wrap (larger gap mesh works best like vegetable and fruit bags) will give the artifact structural support as it undergoes treatment but will not interfere with the introduction of consolidants and stabilizing agents.

An ethulose, PEG, glycerol mixture both strengthens and consolidates these materials, without the adverse swelling associated with using only PEG. It will leave the artifacts natural looking and dry to the touch. Treatment involves placing

[2] There is an unfortunate tendency in conservation to cling to certain developing treatments as perhaps the cure-all for all materials, a magic bullet that will take away the complication of a multiplicity of treatments. Vacuum freeze-drying was one such cure-all and PEG was another. PEG is NOT always a good solution and should be avoided for paper or woven organic materials like grass or pine needles. PEG causes these materials to swell, loosening much of the weaving as the material expands. Silicon Oil is a promising consolidant in the treatment of organics allowing them free motion after treatment and imparting strength to the object. However the treatment is not reversible and its effects have not been observed over prolonged time periods (Smith, 2003).

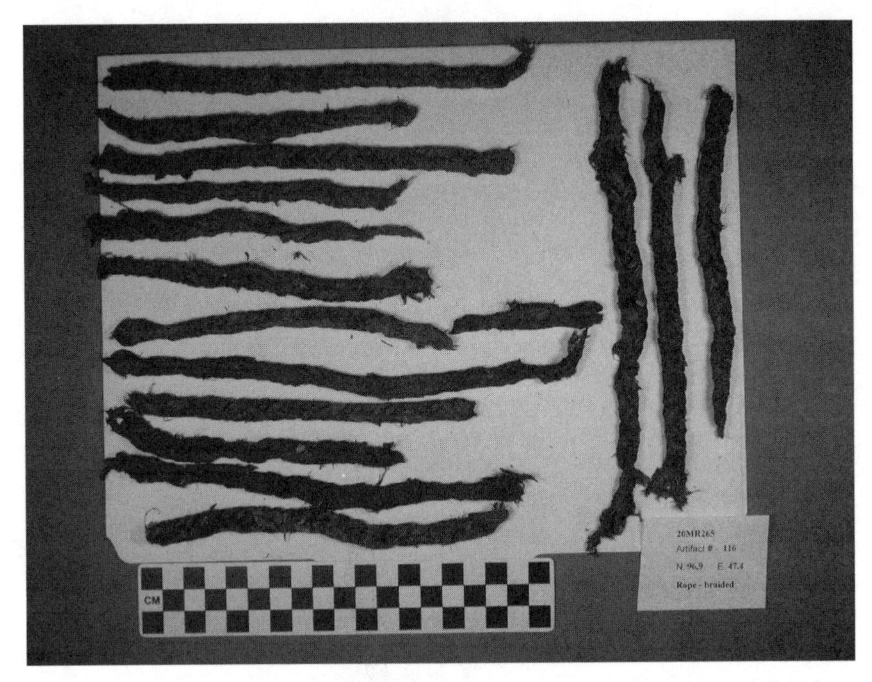

Figure 41. Treated braided rope from an early 19th century site. Photograph by Frank Cantelas.

the cleaned material to be conserved into a solution of 8% PEG 400, .7% ethulose 400, and 2% glycerol (Pearson, 1987: 152). Prior to treatment the artifact should be arranged in a shape for display, as it will remain in this position when treatment is complete.

The desired consolidant mixture is produced by mixing 7 grams of ethulose, 80 grams of PEG 400, and 20 grams of glycerol in 893 grams of distilled water. The ethulose 400 is a powder that should be added to the solution in very small amounts as it absorbs a tremendous amount of water and will clump if it is not mixed thoroughly. This mixture will produce a thick liquid in which the artifact is submerged. It should then be covered, and placed in the refrigerator for six weeks.

DEHYDRATION—METHODOLOGY

After six weeks the cordage or vegetable matting can be removed from the solution, loosely covered and placed in a freezer for atmospheric freeze-drying. Experience has demonstrated that even better results are obtained from placing the artifact on its support in the relative humidity chamber just as it is set up for slow drying wooden artifacts.[3] In eight weeks the water will have sublimated from the

[3] Freezer burn or a slight darkening of the outer layer of an artifact can be a problem in freeze-drying.

cordage and vegetable matting in the freezer. Slow drying will take about half this time, producing an artifact ready for storage or display in four weeks. Care should be taken in the relative humidity chamber that mold does not begin to grow on the artifact or anywhere in the chamber. Spray applications of Lysol will prevent biotic growth in the humidity chamber and should be used as a prophylactic guard on a weekly basis while the equipment is in use.

After dehydration the cordage or vegetable matting can be stored in a climatically controlled environment of about a 50% relative humidity. This humidity level is easily maintained in most air-conditioned buildings. If cardboard boxes are used for storage the artifact must be sealed in a plastic container and checked periodically for mold.

TREATMENT FOR LEATHER—METHODOLOGY

After leather has been cleaned and freed of any stains it is ready for treatment. This treatment is recommended for dry recovered leather and waterlogged leather. Leather is of course made of animal hide, but skin doesn't normally survive well in the archaeological record unless it has been tanned, or converted to leather. Skin is a fibrous polymer of the protein collagen. Tanning causes complex cross-linkages of the collagen, making leather a much more durable material than skin. All leather created before the middle of the 19th century was vegetable tanned, normally with leached tannic acid from oak bark. After the middle of the 19th century industrially tanned leather was produced through infiltration of the heavy metal chromium (chromium tanned leather is green on edge).

The aim of leather treatment is to prevent the charged collagen fibers from the inside of the leather from permanently bonding to the charged fibers on the outside of the leather, via the contractile force of water as it evaporates. This process will permanently shrink and harden the leather. Archaeological leather should only be fully dried after pretreatment.

Pretreatment consists of placing the leather artifact in a 10% PEG 400 solution. At one-week intervals the PEG percentage is raised by 10% increments until a 30% solution is achieved. Placing the artifact in a hot box (see Chapter 1) during treatment will facilitate PEG penetration. One week after the 30% solution is gained the leather artifact is removed and placed in the relative humidity chamber for drying. Atmospheric freeze-drying under a loose plastic cover is also a possibility though results appear better in the humidity chamber.

After a month in the relative humidity chamber or six to eight weeks in the freezer the leather can be removed and evaluated for consolidation or oil application. If necessary loose surface pieces can be consolidated with a brush application of the glue B-72 diluted in toluene.

At this stage the leather should be in good condition though light colored, stiff and dry. This is due to the fact that deterioration has robbed the leather of much of the oil that once kept it supple. Several leather dressings are available to help restore the oil content. One good conditioner currently available is Lexol. Lexol

Figure 42. Early 19th century leather shoe before conditioner application. Photograph by Frank Cantelas.

or other conditioners can be applied with a cloth, paintbrush, or atomizer. Several applications may be necessary to bring back the leather's suppleness and color. If necessary the artifact can be immersed in the conditioner for lengthy soaks. When removed excess conditioner should be daubed off the artifact with a clean cloth.

It should be noted that many of these conditioners will promote corrosion of any metallic fasteners or rivets in the leather. If there are metal staples, rivets, nails, or buttons in the leather the artifact is a composite and will require a slightly different treatment (see Chapter 8).

TREATMENT FOR BONE, TEETH, IVORY, AND ANTLERS—METHODOLOGY

Bone, teeth, antlers, and ivory can be very durable materials provided they have been interred under the right circumstances. They do not last long on the surface of the ground or on the unprotected seabed. All of these materials contain calcium and other mineral salts and are produced in a porous collagen matrix structure, making them susceptible to salt and contaminant intrusion. Since dry finds have been subjected to years of rain percolation and salt intrusion they are treated in the same manner as salt water finds.

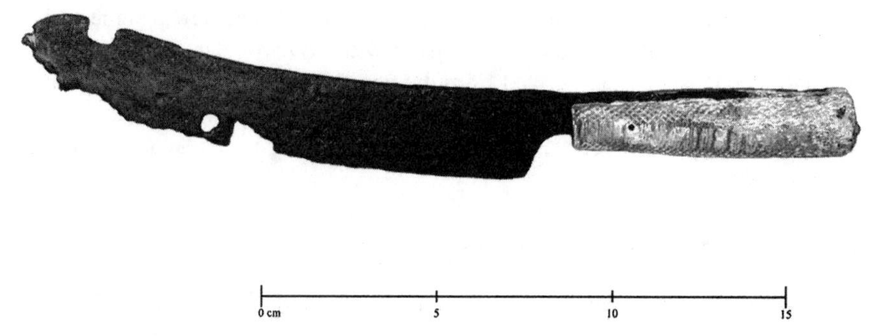

Figure 43. Bone handled knife treated in separate pieces. Photograph by Frank Cantelas.

Basic treatment for these materials consists of a lengthy desalination rinse that can begin when the artifact undergoes stain removal and washing. They should be kept wet in distilled water and stored in the refrigerator between treatments. Following mechanical cleaning and stain removal artifacts of bone, teeth, antlers and ivory, should undergo rinsing similar to ceramics. Agitation rinsing is preferable to static rinsing in this case because of the porosity of the artifacts. This rinse should take up to four weeks with water changes each week.

On completion the artifacts are removed from the rinsing tank and air dried. Freezing, heating, or solvent drying these materials can injure them and may rinse away some of their original fats or oils and should not be contemplated.

If the artifacts are found after desalination to be too delicate to survive dehydration they can be placed in a 50% PVA solution in distilled water and allowed to soak for two weeks. Many white-glues contain PVA formulas and can be substituted for pure PVA. A two-week soak will allow the solution to penetrate the structure of the artifact.

Dehydration should take place in a vapor chamber to keep the glue from being drawn to the surface of the object. A vapor chamber can be any container that can be sealed. Inside, the artifact should be placed on a raised platform and the bottom should be covered with water. A lettuce crisper makes a good vapor chamber. A week of drying should be sufficient in the vapor chamber.

RUBBER

Very little is currently known about rubber conservation. Its use as a waterproofing substance dates to the late 18th century and its more frequent use after vulcanization was perfected in 1839 indicate that it may become a frequent survivor in the 19th century archaeological record. During the early days of its use rubber was processed for use in hoses, shoes, raincoats, and hats. Spreading rubber on cloth to make a waterproof material is how rain gear was fabricated.

Rubber is an organic compound synthesized from latex, the sap of the rubber tree. Latex begins as a colloidal suspension with no particular organization. Its strength and elasticity are derived from the general fiber direction manufactured into it during the refining process. Rubber naturally oxidizes and breaks down over time. Its decomposition is accelerated in direct sunlight or ultraviolet rays and heat.

STORAGE—METHODOLOGY

It seems possible that archaeological sites hold the key to discovering how to preserve this substance for extended periods. Dark, cold, anaerobic conditions have preserved some intact rubber cloth specimens for 140 years or more. This duration is far better than any scientifically developed synthetic anti-oxidants can yet claim.

After recovery rubber should be refrigerated or preferably frozen for storage. The cold will keep the natural breakdown process slowed to a minimum. Preparation for freezing includes placing it in an airtight plastic or non-acidic container flooded with nitrogen to eliminate the possibility of oxidation.

Unfortunately the first indication of rubber breakdown is when it becomes soft and sticky, much like its original latex state. Rubber breakdown is progressive and cannot be halted short of freezing it.

MECHANICAL CLEANING—METHODOLOGY

Rubber artifacts can be cleaned in similar fashion to cellulosic textiles. Liberal use of running water and gentle rubbing should remove most silt and mud. Running water can also help to unfold rubber-impregnated textiles whose edges and corners are too weak to sustain the weight of unfolding them.

Stain removal with EDTA (for metallic salt stains) and hydrogen peroxide (for sulfide stains) does work but may have unexpected consequences at a later time. Thus far very little rubber has been treated and stored.

TREATMENT FOR RUBBER—METHODOLOGY

There is no currently recommended rubber conservation treatment. PEG 400 has been applied via atomizer, as have various commercially available rubber antioxidants. Though these products invariably make the rubber cloth more flexible and natural in texture and color, they have yet to be tested over long periods.

Until a treatment is developed and tested for long-term results, rubber artifacts should be refrigerated or frozen. Larger artifacts can be stored on acid free rolls inside plastic tubes filled with an inert gas such as nitrogen (see Storage and Recovery).

CONCLUSION

Conservation of organic artifacts other than wood offers a great challenge to the archaeologist/conservator. No other category of materials demand so much time, hands on treatment, and scrutiny as do organic artifacts. Nor are the risks of a miss-step as great with other material types as they are with delicate organics.

Yet the rewards in micro-excavation and conservation of organic materials can also exceed those of other materials in relating history and usage. One piece of well preserved and conserved tartan cloth can help identify a tribe or a civilization better than virtually any other artifact. And though it may seem counterintuitive that these delicate organic artifacts survive in the archaeological record, they do indeed, time and again.

CHAPTER 7: ORGANICS OTHER THAN WOOD

Ambrose, W.R. and M.J. Mummery. "The Use of Aqueous Glycerol Solutions in the Conservation of Waterlogged Archaeological Organic Materials" *Archaeometry: Further Australasian Studies* (Canberra) (1987): 292–302.

Anderson, Adrienne. "The Archaeology of Mass-Produced Footwear." *Historical Archaeology 1968. Society for Historical Archaeology*, Vol. 2, pp. 56–65. [ECU-525]

Andrews, Theresa Meyer, William W. Andrews and Cathleen Baker. "An Investigation into the Removal of Enzymes from Paper Following Conservation Treatment." *Journal of the American Institute for Conservation* Vol. 31 No. 3, 1992. [ECU-724]

Anon. "A Comparitive Investigation of Methods for the Consolidation of Wet Archaeological Leather: Application of Freeze-Drying to Polyethylene Glycol Impregnated Leather," *Icom Proceedings*, Symposium of Ethnographic and Waterlogged Leather CL Publication, Amsterdam (1986): 66–69.

Anon. "*Wasa's* Sails Unfurled," *New Scientist* 66: 953 (1975)L 669. [ECU-274]

Arnaud, G., S. Arnaud, A. Ascenzi, E. Bonucci, and G. Graziani. "On the Problem of the Preservation of Human Bone in Sea Water." *International Journal of Nautical Archaeology* 9:1 (1980): 53–63. [ECU-168]

The Art of the Ancient Weaver Textiles from Egypt (4th–12th century A.D.) Ann Arbor, MI: Kelsey Museum of Archaeology and the University of Michigan, 1980. [ECU-582]

Baer, S.V.S. and N. Indictor. "Organic Chemistry for Conservation Students." *Studies in Conservation* 20:1 (Book Review)(1975): 40–42.

Barandiaran, Marta. "Evaluation of Conservation Treatments Applied to Salted Paper Prints, Cyanotypes, and Platinotypes." *Studies in Conservation* Vol. 45 No. 3, 2000, pp. 162–168. [ECU-725]

Beardmore, J. and V. Jenssen. "The Recovery and Conservation of Waterlogged Grass Matting Fragments from a Marine Enviroment." (Unpublished Report, Materials Laboratory, Conservation Division, Parks Canada, Ottawa) (1981).

Bengtsson, Sven. "Preservation of the *Wasa* Sails." *Conservation in Art and Archaeology* (London, IIC) (1975): 33–35.

—. "The Sails of the *Wasa*: Unfolding, Identification, and Preservation." *International Journal of Nautical Archaeology* 4 (1975): 27–41. [ECU-167]

Braceckle, Irene, Jonathan Thornton, Kimberly Nichols, and Gerri Strickler. "Cyclododecane: Technical Note on Some Uses in Paper and Object Conservation." *Journal of the American Institute for Conservation* Vol. 38 No. 2, 1999. [ECU-726]

Brady, P.R., G.N. Freeland, R.J. Hine, and R.M. Hoskinson. "The Absorption of Certain Metal Ions by Wool Fibers." *Textile Research Journal* 44 (1974): 733–735. [ECU-216]

Breheny, E. "Conservation of a Tin Glazed Earthen Ware Relief." *Scottish Society for Conservation and Restoration.* Vol. 14 No. 2, 1993. [ECU-789]

Brooke, S. and K. Morris. "A Preliminary Report on the Conservation of Marine Archaeological Artifacts from the *Defense*." (Bulletin Preprints, American Institute for Conservation, Boston) (1977): 7–34.

Brothwell, Don R. *Digging up Bones*. London: British Museum of Natural History, 1972.

Brown, Margaret K. "A Preservative Compound for Archaeological Materials." *American Antiquity* 39:3 (1974): 469–473.

Calnan, C.N. "An Investigation into the Use of Chelating Agents in the Treatment of Iron Stains on Waterlogged Leather." (Unpublished Thesis, Queens University, Ontario, Canada) (1976).

Cardamone, Carter I.M. "Evaluation of Degradation in Museum Textiles using Infrared Photography and Property Kinetics." (University of Minnesota Publication) (1983).

CCI Laboratory Staff. "Anionic Detergent." *CCI Notes* (Canadian Conservation Institute) 13/9 (1986). [ECU-462]

CCI Laboratory Staff. "Fibre Information." *CCI Notes* (Canadian Conservation Institute) 13/11 (1986). [ECU-445]

CCI Laboratory Staff. "Washing of Non-Coloured Textiles." *CCI Notes* (Canadian Conservation Institute) 13/7 (1986). [ECU-448]

CCI Laboratory Staff. "Cleaning Paintings: Precautions" *CCI Notes* (Canadian Conservation Institute) 10/1 (1995). [ECU-538]

CCI Laboratory Staff. "Storing Works on Paper." *CCI Notes*. (Canadian Conservation Institute) 11/2 (1995). [ECU-540]

CCI Laboratory Staff. "Basic Care of Books." *CCI Notes*. (Canadian Conservation Institute) 11/7 (1995). [ECU-541]

CCI Laboratory Staff. "Framing Works of Art on Paper." *CCI Notes* (Canadian Conservation Institute) 11/9 (1995). [ECU-542]

CCI Laboratory Staff. "Mounting Small, Light, Flat Textiles" *CCI Notes* (Canadian Conservation Institute) 13/6 (1995). [ECU-544]

CCI Laboratory Staff. "Care of Encased Photographic Images" *CCI Notes* (Canadian Conservation Institute) 16/1 (1995). [ECU-549]

CCI Laboratory Staff. "Applying Accession Numbers to Textiles." *CCI Notes* (Canadian Conservation Institute) 13/8 (1986). [ECU-446]

CCI Laboratory Staff. "Care of Ivory, Bone, Horn, and Antler." *CCI Notes* (Canadian Conservation Institute) 6/1 (1988). [ECU-470]

CCI Laboratory Staff. "Mounting Small, Light, Flat Textiles." *CCI Notes* (Canadian Conservation Institute) 13/6 (1988). [ECU-447]

CCI Laboratory Staff. "Stitches Used in Textile Conservation." *CCI Notes* (Canadian Conservation Institute) 13/10 (1988). [ECU-444]

CCI Laboratory Staff. "Care of Objects Made from Rubber and Plastic." *CCI Notes* (Canadian Conservation Institute) 15/1 (1988). [ECU-440]

CCI Laboratory Staff. "Care of Encased Photographic Images." *CCI Notes* (Canadian Conservation Institute) 16/1 (1988). [ECU-441]

CCI Laboratory Staff. "Commercial Dry Cleaning of Museum Textiles." *CCI Notes* (Canadian Conservation Institute) 13/13 (1990). [ECU-372]

CCI Laboratory Staff. "Velcro Support System for Textiles." *CCI Notes* (Canadian Conservation Institute) 13/4 (1991). [ECU-356]

CCI Laboratory Staff. "Care of Alum, Vegetable, and Mineral Tanned Leather." *CCI Notes* (Canadian Conservation Institute) 8/2 (1992). [ECU-456]

CCI Laboratory Staff. "Care of Rawhide and Semi-Tanned Leather." *CCI Notes* (Canadian Conservation Institute) 8/4 (1992). [ECU-468]

CCI Laboratory Staff. "Textiles and the Environment." *CCI Notes* (Canadian Conservation Institute) 13/1 (1992). [ECU-457]

CCI Laboratory Staff. "Removing Mould from Leather." *CCI Notes* (Canadian Conservation Institute) 8/1 (1993). [ECU-469]

CCI Laboratory Staff. "Storage and Display Guidelines for Paintings." *CCI Notes* (Canadian Conservation Institute) 10/3 (1993). [ECU-504]

CCI Laboratory Staff. "Environmental and Display Guidelines for Paintings." *CCI Notes* (Canadian Conservation Institute) 10/4 (1993). [ECU-474]

CCI Laboratory Staff. "Stitches Used in Textile Conservation." *CCI Notes* (Canadian Conservation Institute) 13/10 (1995). [ECU-545]

CCI Laboratory Staff. "Commercial Dry Cleaning of Museum Textiles." *CCI Notes* (Canadian Conservation Institute) 31/13 (1995). [ECU-546]

CCI Laboratory Staff. "Condition Reporting—Paintings. Part I: Introduction." *CCI Notes* (Canadian Conservation Institute) 10/6 (1993). [ECU-501]

CCI Laboratory Staff. "Condition Reporting—Paintings. Part II: Examination Techniques and a Checklist." *CCI Notes* (Canadian Conservation Institute 10/7) (1993). [ECU-502]

CCI Laboratory Staff. "Framing a Painting." *CCI Notes* (Canadian Conservation Institute) 10/8 (1993). [ECU-503]

CCI Laboratory Staff. "Keying Out of Paintings." *CCI Notes* (Canadian Conservation Institute) 10/9 (1993). [ECU-505]

CCI Laboratory Staff. "Backing Boards for Paintings on Canvas." *CCI Notes* (Canadian Conservation Institute) 10/10 (1993) [ECU-475]

CCI Laboratory Staff. "Removing a Painting from Its Frame." *CCI Notes* (Canadian Conservation Institute) 10/12 (1993). [ECU-506]

CCI Laboratory Staff. "Basic Handling of Paintings." *CCI Notes* (Canadian Conservation Institute) 10/13 (1993). [ECU-476]

CCI Laboratory Staff. "Care of Paintings on Ivory, Metal, and Glass." *CCI Notes* (Canadian Conservation Institute) 10/14 (1993). [ECU-477]

CCI Laboratory Staff. "Paintings: Considerations Prior to Travel." *CCI Notes* (Canadian Conservation Institute) 10/15 (1993). [ECU-478]

CCI Laboratory Staff. "Wheat Starch Paste." *CCI Notes* (Canadian Conservation Institute) 11/4 (1993). [ECU-479]

CCI Laboratory Staff. "Removing Paper Artifacts from Their Frames." *CCI Notes* (Canadian Conservation Institute) (1993). [ECU-500]

CCI Laboratory Staff. "Flat Storage for Textiles." *CCI Notes* (Canadian Conservation Institute) 13/2 (1993). [ECU-464]

CCI Laboratory Staff. "Rolled Storage for Textiles." *CCI Notes* (Canadian Conservation Institute) 13/3 (1993). [ECU-449]

CCI Laboratory Staff. "Washing Non-coloured Textiles." *CCI Notes* (Canadian Conservation Institute) 13/7 (1993). [ECU-480]

CCI Laboratory Staff. "The Beilstein Test: Screening Organic and Polymeric Materials for the Presence of Chlorine, with Examples of Products Tested." *CCI Notes* (Canadian Conservation Institute) 17/1 (1993). [ECU-507]

CCI Laboratory Staff. "Making a Mini Vacuum Cleaner." *CCI Notes* (Canadian Conservation Institute) 18/2 (1993). [ECU-508]

CCI Laboratory Staff. "Care of Objects Decorated with Glass Beads." *CCI Notes* (Canadian Conservation Institute) 6/4 (1994). [ECU-510]

CCI Laboratory Staff. "Condition Reporting—Paintings Part III: Glossary of Terms." *CCI Notes* (Canadian Conservation Institute) 10/11 (1994). [ECU-511]

CCI Laboratory Staff. "Display Methods for Books." *CCI Notes* (Canadian Conservation Institute) 11/8 (1994). [ECU-512]

CCI Laboratory Staff. "Hanging Storage for Costumes." *CCI Notes* (Canadian Conservation Institute) 13/5 (1994). [ECU-513]

CCI Laboratory Staff. "Storage for Costume Accessories." *CCI Notes* (Canadian Conservation Institute) 13/12 (1994). [ECU-514]

CCI Laboratory Staff. "Display and Storage of Museum Objects Containing Cellulose Nitrate." *CCI Notes* (Canadian Conservation Institute 15/3) (1994). [ECU-515]

CCI Laboratory Staff. "Applying Accession Numbers to Textiles." *CCI Notes* (Canadian Conservation Institute) 13/8 (1994). [ECU-516]

Clark, Susie. "The Conservation of Wet Collodion Positives." *Studies in Conservation* Vol. 43 No. 4, 1998, pp. 231–241 [ECU-727]

Clavir, Miriam. "An Initial Approach to the Stabilization of Rubber from Archaeological Sites and in Museum Collections." *Journal of the IIC* (Canadian Group 7) (1982): 1–10. [ECU-430]

Coles, John M. "An Assembly of Death: Bog Bodies of Northern and Western Europe." *Wet Site Archaeology* B.A. Purdy, ed. (1988): 219–235.

"A Comparative Investigation of Methods for the Consolidation of Wet Archaeological Leather: Application of Freeze-Drying to Polyethylene Glycol Impregnated Leather." *ICOM Proceedings* (Symposium of Ethnographic and Waterlogged Leather CL Publication, Amsterdam) (1986): 66–69.

Cook, J. Gordon. *Handbook of Textile Fibers*. Watford, U.K.: Marrow, 1959. [ECU-488]

Cowan, Janet. "Dry Methods for Surface Cleaning Paper." *CCI Technical Bulletin* 11 (Canadian Conservation Institute) (1986). [ECU-451]

Croucher, R. and A.R. Wooley. *Fossils, Minerals and Rocks: Their Collection and Preservation*. Cambridge: Cambridge University Press for the British Museum of Natural History, 1983.

David, A. "Freeze-Drying Leather with Glycerol." *Museums Journal* 81:2 (1981): 103–104.

Denton, M.H. and J.S. Gardner. "The Recovery and Conservation of Waterlogged Goods Excavated at the Fort Loudoun Site, Fort Loudoun, Pennsylvania." *Historical Archaeology* 17:1 (1983): 96–103.

Dickinson, H.W. "A Condensed History of Rope Making." *American Antiquity* 39:3 (1974): 469–473.

Dorrego, F., M.P. Luxan, and M. Ruiz. "Reactivity of Natural Fat with Different Compounds in Historical Surface Coatings." *Surface Coatings International* Vol. 77 No. 2, 1994. [ECU-616].

Edge, P. "Textiles from the *Mary Rose*." *UKIC Conservation News* 18 (1982): 13–14.

"Effects of Aqueous Exposure on the Mechanical Properties of Wool Fibers—Analysis by Atomic Force Microscopy." *Textile Research Journal* Vol. 71 No. 7, 2001, pp. 573–581. [ECU790]

Ellam, Diane. "Wet Bone: The Potential for Freeze-Drying." *Archaeological Bone, Antler and Ivory* (Occasional Papers No. 5, K. Starling and D. Watkinson, eds., London: United Kingdom Institute for Conservation) (1987): 34–35.

"Experiments in the Conservation of Rope." (Unpublished Maunscript) [n.d.] [ECU-0484]

Feller, Robert L. "Fundamentals of Conservation Science: Induction Time and the Auto Oxidation of Organic Compounds." *Bulletin of the American Institute for Conservation* 14:2 (1974): 142–151.

Feniak, Christine. "Removal of Rust Staining from Waterlogged Silk Fragments." (Unpublished Manuscript) (1979). [ECU-483]

Florian, Mary-Lou, E., D.P. Kronkright, and R.E. Norton. *The Conservation of Artifacts made from Plant Materials*. Marina Del Ray Getty Conservation Institute, 1990.

Fogle, S., ed. *Recent Advances in Leather Conservation*. (The Foundation of the American Institute for Conservation of Historic and Artistic Works, Washington D.C.) (1985).

Frame, M. "Conservation of Archaeological Textiles." *Textile Conservation Newsletter* (1982): 6–7.

Fraser, R.D.B. "Keratin." *Scientific American* (1969): 87–96. [ECU-0267]

Fraser, O. "An Examination of Waterlogged Cremated Bone." *Scottish Society for Conservation and Restoration* Vol. 5 No. 3, 1994. [ECU-791]

Fry, M.F. "An Economy Vacuum Freeze-Dryer for Archaeological Organic Materials." *Vacuum* 5 (1984): 555–558.

Ganiaris, H., Keene, S., and K. Starling. "A Comparison of Some Treatments for Excavated Leather." *The Conservator* 6 (1982): 12–23. [ECU-560]

Garstang, J.H. "Document Analysis." (Engineering Branch Report, Transportation Safety Board, Canada) (1989). [ECU-417]

Godfrey, I.M., E.L. Ghisalberti, E.W. Beng, L.T. Byrne and G.W. Richardson. "The Analysis of Ivory from a Marine Environment." *Studies in Conservation* Vol. 47 No. 1, 2002, pp. 29–45. [ECU-728]

Gregory, David. "Experiments into the Deterioration Characteristics of Materials on the Duart Point Wreck Site: An Interim Report." *International Journal of Nautical Archaeology* Vol. 24 No. 1, 1995, pp. 61–65. [ECU730]

Gross, J. "Collagen." *Scientific American* 204 (1961): 121–130. [ECU-0268]

Grosso, Gerald H. "Wood, Textile and Leather Conservation Techniques for the Archaeologist." *Northwest Anthropological Research* Notes 9:1 (1975): 180–197.

Hallebeek, P.B., ed. *Symposium on Ethnographic and Waterlogged Leather* (ICOM Committee for Conservation, Amsterdam) (1986).

Halverson, Bonnie G. and Nancy Kerr. "The Effects of Light on the Properties of Site Faricx Coated with Parylene—C." *Studies in Conservation* Vol. 39 No. 1 (Feb. 1994). [ECU-558]

Hamilton, Donny L, and C.Wayne Smith. "Use of MTMS in the Preservation of Thin-Section Tissue Samples." *TAMUS #1756—Technology Licensing Bulletin*, 2001. [ECU-792]

Harris, Milton. *Handbook of Textile Fibers* (Harris Research Laboratories, Washington, D.C.) (1954). [ECU-487]

Hawley, Janet K. "The Conservation of Waterlogged Rope from a Sixteenth Century Basque Whaling Ship." *ICOM Proceedings* (Fremantle) (1987): 19–37. [ECU-196]

Hawley, J.K., E.A. Kawai, and C. Sergeant. "The Removal of Rust Stains from Arctic Tin Can Labels Using Sodium Hydrosulfite." *Journal of the IIC-Canadian Group* 6 (1 & 2) (1981): 17–24.

Heard, Teresa A. and Sara J. Kadolph. "Technical Report: Storage Practices in Textile and Costume Collections." (Unpublished Manuscript) (1993). [ECU-463]

Heil, Adolph, and W. Esch. *The Manufacture of Rubber Goods*. London: Charles Griffin, 1923.

Henderson, G. and M. Stanburg. "The Excavation of a Collection of Cordage from a Shipwreck Site." *International Journal of Nautical Archaeology* 12:1 (1983): 15–36.

Hillman, David. "A Short History of Early Consumer Plastics." *Journal of the Canadian Conservation Institute* No.10&11, (1985): 20–27.

Hofenk de Graaff, J.H. "The Constitution of Detergents in Connection with the Cleaning of Ancient Textiles." *Studies in Conservation* 13 (1968): 122–141. [ECU-336]

Hoffman, Per. Proceedings of the 4th ICOM Group on Wet Organic Archaeologcial Materials. Bremerhaver, 1991. [ECU-431]

Howell, David. "Ageing and Degradation of Textiles." *CCI Newsletter* (Canadian Conservation Institute) No. 21, 1998. [ECU-731]

Hudson, Julie. "The Weaver's Sketchbook." *Smoke Fire News* Vol. 5 No. 11, August 1991 (reprint). [ECU-605]

Jenssen, Victoria. "Water-Degraded Organic Materials: Skeletons in Our Closets?" *Museum* 35:1 (1983): 15–21. [ECU-498]

—. "Conservation of Wet Organic Artefacts Excluding Wood." [ECU-559]

Koob, Stephen P. "The Consolidation of Archaeological Bone." *Adhesives and Consolidants* N. Bromellie, E. Pye, D. Smith, and G. Thomson, eds.(Contributions to the 1984 Paris Congress, International Institute for Conservation of Historic and Artistic Works) (1984): 98–102.

Laroche, Cheryl and Gary S. McGowan. "'Material Culture': Conservation and Analysis of Textile Recovered from Five Points." *Historical Archaeology* Vol. 35 No. 3, 2001. [ECU-732]

Logan, Judith A. "Red Bay 1982—Textile Discovery." *Textile Conservation Newsletter* (Canada) (1983): 3–5. [BCU-03621

Luniak, Bruno. *The Identification of Textile Fibres: Qualitative and Quantitative Analysis of Fibre Blends* Sir Isaac Pitman & Sons, Ltd. [n.d.]. [ECU-493]

MacLeod, Ian D., Reid, N.K.M., and N. Sander. "Conservation of Waterlogged Organic Materials: Comments on the Analysis of PEG and the Treatment of Leather and Rope." *ICOM Proceedings* (7th Triennial Meeting, Copenhagen) (1984).

Maltby, Susan L. "Rubber, the Problem that Becomes a Solution." *Studies in Conservation* (1988).

Mason, J. "Report on Waterlogged Leather Experiment: December 1980–January 1981." (Canadian Conservation Institute, Archaeology Division) Report No. 398 (1981).

McHugh, Maureen Colin. "How to Wet-Clean Undyed Cotton and Linen." (Leaflet, Smithsonian Institution Museum of History and Technology Textile Laboratory, Washington, D.C.) (1967). [ECU-0492]

"A Method for the Determination of Aerobic Biodegrability." *Process Biochemistry* Vol. 28 No. 2, 1993. [ECU-793]

Mills, John S. and Raymond White. *Organic Chemistry of Museum Objects*. London: Elsevier, 1999.

Morris, K. and B. Seifert. "Conservation of Leather and Textiles from the *Defense*." *Journal of the American Institute for Conservation* 18:1 (1978): 33–43. [ECU-482]

Morse, Elizabeth. "Enzyme Treatments for Conserving Artistic/Historic Works: A Selected Bibliography—1940–1990." *Technology and Conservation* (Spring 1992): 20–24. [ECU-426]

Muhlethaler, B., Barkman, L., and D. Noack. *Conservation of Waterlogged Wood and Wet Leather*. Paris: Edition Eyrothes, 1973. [ECU-369]

Murray, H. "The Conservation of Organic Artifacts from the *Mary Rose*." *Science Technology* 7 (1984): 19–24.

—. "The Use of Mannitol in Freeze-Drying Waterlogged Organic Material." *The Conservator* 9 (1985): 33–35. [ECU-411]

Nockert, M. and T. Wadsten. "Storage of Archaeological Textile Finds in Sealed Boxes." *Studies in Conservation* 23 (1978): 38–41. [ECU-185]

O'Connor, Sonia. "The Identification of Osseous and Keratinous Materials at York." *Archaeological Bone, Antler, and Ivory* K. Starling and D. Watkinson, eds.(Occasional Papers No. 5, London: United Kingdom Institute for Conservation) (1987): 9–21.

O'Connor, Terry P. "On the Structure, Chemistry, and Decay of Bone, Antler, and Ivory." *Archaeological Bone, Antler, and Ivory* K. Starling and D. Watkinson, eds. (Occasional Papers No. 5, London: United Kingdom Institute for Conservation) (1987): 6–8.

0'Floinn, Raghnall. "Irish Bog Bodies." *Archaeology Ireland 2* (1988): 94–97.

O'Shea, C. "The Use of Dewatering Fluids in the Conservation of Waterlogged Wood and Leather." *Museums Journal* 71:2 (1971): 71–72. [ECU-413]

Panter, I. "The Conservation of Leather Artifacts Recovered from the *Mary Rose*." *Journal of the Society of Leather Technologists and Chemists* (1982): 30–31. [ECU-114]

—. "An Investigation into Improved Methods for the Conservation of Waterlogged Leather." *Bulletin* (Scottish Society for Conservation and Restoration 7) (1986): 2–11.

"Parylene Polymer Conserves Historical Material." (Advertisement, Nova Tran Corporation, Clear Lake, Wisconsin) [n.d.]. [ECU-416]

Pate, F. Donald. "Bone Collagen Preservation at the Roonka Flat Aboriginal Burial Ground: A Natural Laboratory." *Journal of Field Archaeology* Vol. 25 No. 2, 1998, pp. 203–217. [ECU-733]

Paterakis, Antonio. "Conservation of a Late Minoan Basket from Crete." *Studies in Conservation* Vol. 41 No. 3, 1996, pp. 179–182. [ECU-734]

Pearson, Colin, ed. *Conservation of Marine Archaeological Objects*. London, Butterworth Series in Conservation and Museology, 1987.

Peacock, Elizabeth E. "Deacidification of Degraded Linen." *Studies in Conservation* 28 (1983): 8–14. [ECU-121]

—. "The Conservation and Restoration of Some Anglo-Scandinavian Leather Shoes." *The Conservator* 7 (1983): 18–23.

Pearson, Herbert. *Waterproofing Textile Fabric*. New York: Chemical Catalogue Company, 1924.

Peterman, Glen. "Papyrus Scrolls from Petra: A Stupendous Discovery." *Biblical Archaeologist* Vol. 54 No. 1, 1994.

Quye, Anita. "Historical Plastics Come of Age." *Chemistry in Britain* Vol. 31 No. 8, 1995, pp. 617–620. [ECU-736]

Rector, W.K. "The Treatment of Waterlogged Leather from Blackfriars, City of London." (Transactions of the Museum Assistants Group 12) (1975): 3337.

Reed, R. *Science for Students of Leather Technology*. London: Pergamon Press, (1966).

—. *Ancient Skins, Parchments, and Leather*. London: Seminar Press, 1972.

Rixon, A.E. *Fossil Animal Remains: Their Preparation and Conservation*. London: Athione, 1976.

Rodgers, Bradley A. *The East Carolina University Conservator's Cookbook: A Methodological Approach to the Conservation of Water Soaked Artifacts*. A Herbert R. Paschal Memorial Fund Publication, East Carolina University, Program in Maritime History and Underwater Research, 1992. [ECU-402]

Rosenquist, A.M. "Experiments on the Conservation of Waterlogged Wood and Leather by Freeze-Drying." *Maritime Monographs and Reports* (National Maritime Museum, Greenwich, U.K.) 16 (1975): 9–23. [ECU-258]

Santappa, M. and Rao V.S. Sundara. "Vegetable Tannins—A Review." *Journal of Scientific and Industrial Research 41* (1982): 705–718. [ECU-392]

Sease, C. "The Case Against Using Soluable Nylon in Conservation Work." *Studies in Conservation* 26:3 (1981): 102–110. [ECU-397]

—. *A Conservation Manual for the Field Archaeologist*. (Archaeological Research Tools Vol. 4, Institute of Archaeology, University of California, Los Angeles) (1987). [ECU-370]

Segal, Martha. "Treatment of the Textiles from Red Bay, Labrador." *Textile Conservation Newsletter* (Canada) (1983): 5–7. [ECU-361]

Sills, Bernard and Seymour Couzyn. "Dry Preservation of Biologic Specimens by Plastic Infiltration." *Curator* 4:1 (1958): 72–75. [ECU-85]

Siu, Ralph G. *Microbial Decomposition of Cellulose*. New York: Reinhold, 1951.

Smith, C.Wayne and Sylvia Grider. "The Emergency Conservation of Waterlogged Bibles from the Memorabilia Assemblage Following the Collapse of the Texas A and M University Bonfire." *International Journal of Historical Archaeology* Vol. 5 No. 4, 2001. [ECU-794]

Smith, C. Wayne. *Archeological Conservation Using Polymers; Practical Applications for Organic Artifact Stabilization*. College Station, Texas A & M University Press, 2003.

Smith, C.Wayne, Ellen M. Heath, D. Andrews Merriweather, and David Reed. "Placing a Face in History: Excavation Facial Re-Construction and DNA Analysis of Skeletal Remains from LaSalle's Vessel, *LaBelle*." [ECU-795]

Snow, Carol E. and Terry Drayman Weisser. "The Examination and Treatment of Ivory and Related Materials." *Adhesives and Consolidants* N. Brommellie, E. Pye, D. Smith, and G. Thomson, eds. (Contributions to the 1984 Paris Congress, International Institute for Conservation of Historic and Artistic Works) (1984): 141–145.

Stambolov, T. "Manufacture, Deterioration, and Preservation of Leather." (ICOM Survey of Literature of Theoretical Aspects and Ancient Techniques) (1969).

Starling, K. "The Freeze-Drying of Leather Pre-Treated with Glycerol." *Preprints* (ICOM 7th Triennial Meeting, Copenhagen) (1984): 22–25.

Stone, Tammy T., David N. Dickel, and Glen H. Doran. "The Preservation and Conservation of Waterlogged Bone from the Windover Site, Florida: A Comparison of Methods." *Journal of Field Archaeology* 17:2 (1990): 177–186.

Stone, T. "Removing Mould from Leather." *Notes* (Canadian Conservation Institute) 8/1 (1991). [ECU-353]

Strang, T. "Characterization of Waterlogged Archaeological Cork: A Prelude to Treatment." (Unpublished thesis. Queens University, Ontario, Canada) (1984).

Stroz, Michael D., Robert H. Glew, Stephan L. Williams and Asish K. Saha. "Comparisons of Preservation Treatments of Collagen Using the Collagenase—SDS—PAGE Technique." *Studies in Conservation* Vol. 38 No. 1, 1993. [ECU-737]

Sturge, T. "The Conservation of Wet Leather." (Unpublished Thesis. University of London, Institute of Archaeology, England) (1975).

Sturman. "Sorbitol Treatment of Leather and Skin: A Preliminary Report." *Preprints* (ICOM 7th Triennial Meeting, Copenhagen) (1984).

Sugarman, Jane E. and Timoth J. Vitale. "Observations on the Drying of Paper: Five Drying Methods and the Drying Process." *Journal of the American Institute for Conservation* Vol. 31 No. 2, 1992. [ECU-738]

Szczepanowska, Hanna, and Charles M. Lovett. "A Study of the Removal and Prevention of Fungal Stains on Paper." *Journal of the American Institute for Conservation* Vol. 31 No. 2, 1992. [ECU-739]

Tarleton, Kathryn S. and Margaret T. Ordoñez. "Stabilization Methods for Textiles from Wet Sites." *Journal of Field Archeology* Vol. 22 No. 1, 1995, pp. 81–95. [ECU-740]

Technology & Conservation Magazine of Art, Architecture, and Antiquities. Vol. 12 No. 4, Winter 1996. [ECU-603]

Textile Conservation Newsletter. (Canadian Conservation Institute, Ottawa, Ontario, Canada) (1985): 17–19. [ECU-363]

"Textiles in the Kelsey Museum of Archaeology." (Kelsey Museum of Archaeology) [n.d.]. [ECU-427]

Thickett, David, Pipa Cruickshank, and Clare Ward. "The Conservation of Amber." *Studies in Conservation* Vol. 40 No. 4, 1995, pp. 217–226. [ECU-741]

Tilbrooke, D.R.W. "Removal of Iron Contaminants from Wet Leather Using Complexing Agents in Dipolar Aprotic Solvents." *ICOM Proceedings* (5th Triennial meeting, Zagreb) (1978).

Tilbrooke, D.R.W. and Colin Pearson. "The Conservation of Canvas and Rope Recovered from the Sea." *Proceedings* (Pacific Northwest Conference 1) (1976): 61–66. [ECU-235]

Timar-Balazsy, Agnes and Dinih Eastop. *Chemical Principles of Textile Conservation*. London: Butterworths, 1998.

Tse, Season, and Dorit von Derschau. "The Science of Conservation: Surfactant Residue and Rinsing Procedures for Historic Textiles." *CCI Newsletter* (Canadian Conservation Institute No. 27, 2001. [ECU-742]

Van Der Reyden, Dianne. "Recent Scientific Research in Paper Conservation." *Journal of the American Institute for Conservation* Vol. 31 No. 1, 1992. [ECU-743]

Van Dienst, Elise. "Some Remarks on the Conservation of Wet Archaeological Leather." *Studies in Conservation* 30:2 (1985): 86–92. [ECU-309, 561]

Von Soest, H.A.B., T. Stambolov, and P.B. Hallebeek. "Conservation of Leather." *Studies in Conservation* 29:1 (1984): 21–31. [ECU-188]

Vuori, Jan. "Conservation Treatment of the Kanehsatake Flag." *CCI Newsletter* (Canadian Conservation Insitute) No. 27, 2001. [ECU-744]

Vuori, Jan and Robin Hanson. "Conservation of a Military Tunic Including the Use of Guide Threads for Positioning Repairs." *Journal of the American Institute for Conservation* Vol. 39 No. 2, 2000. [ECU-745]

"*Wasa*'s Sails Unfurled." *New Scientist* 66:953 (1975): 669. [ECU-274]

"Wear and Fatigue of Nylon and Polyester Mooring Lines." *Textile Research Journal* Vol 67, No. 7, 1997. [ECU-796]

Webb, Marianne. Lacquer: Technology and Conservation. London: Butterworths, 2000.

Webster, R. "Ivory, Bone, and Horn." *The Gemnologist* 27:322 (1958): 91–98.

—. "Vegetable Ivory and Tortoise Shell." *The Gemnologist* 27:323 (1958): 103–107.

Weindling, Ludwig. *Long Vegetable Fibers*. New York: Columbia University Press, 1947.

Wolock, I. "Polymeric Materials and Composite." *Seawater Corrosion Handbook*. M. Schumacher, ed., Park Ridge, New Jersey: Noyes Data Corporation, 1979: 458–459.

Worden, E.C. *Nitrocellulose Industry*. New York: Van Nostrand, 1911.

Wouters, J. "A Comparative Investigation of Methods for the Consolidation of Wet Archaeological Leather: Application of PEG-Impregnation to a Shoe from the 13th Century." *ICOM Proceedings* (7th Triennial Meeting, Copenhagen) (1984): 29–31.

"X-Ray Diffractometric Measurement of Microcrystallite Size Unit Cell Dimensions and Crystallinity: Application to Cellulosic Marine Textiles." *Textile Research Journal* Vol. 63 No. 8, 1993, pp. 755–464. [ECU-797]

Young, Gregory S. "Analytical Research on the Conservation of Native Skins and Leathers." *CCI Newsletter* (Canadian Conservation Institute) (1988): 6–7. [ECU-100]

Zaitseva, G.A. "Protection of Museum Textiles and Leather Against the Dermestid Beetle (Coleoptera, Dermestidae) by Means of Antifeedants." *Studies in Conservation* 32 (1987): 176–18 [ECU-486]

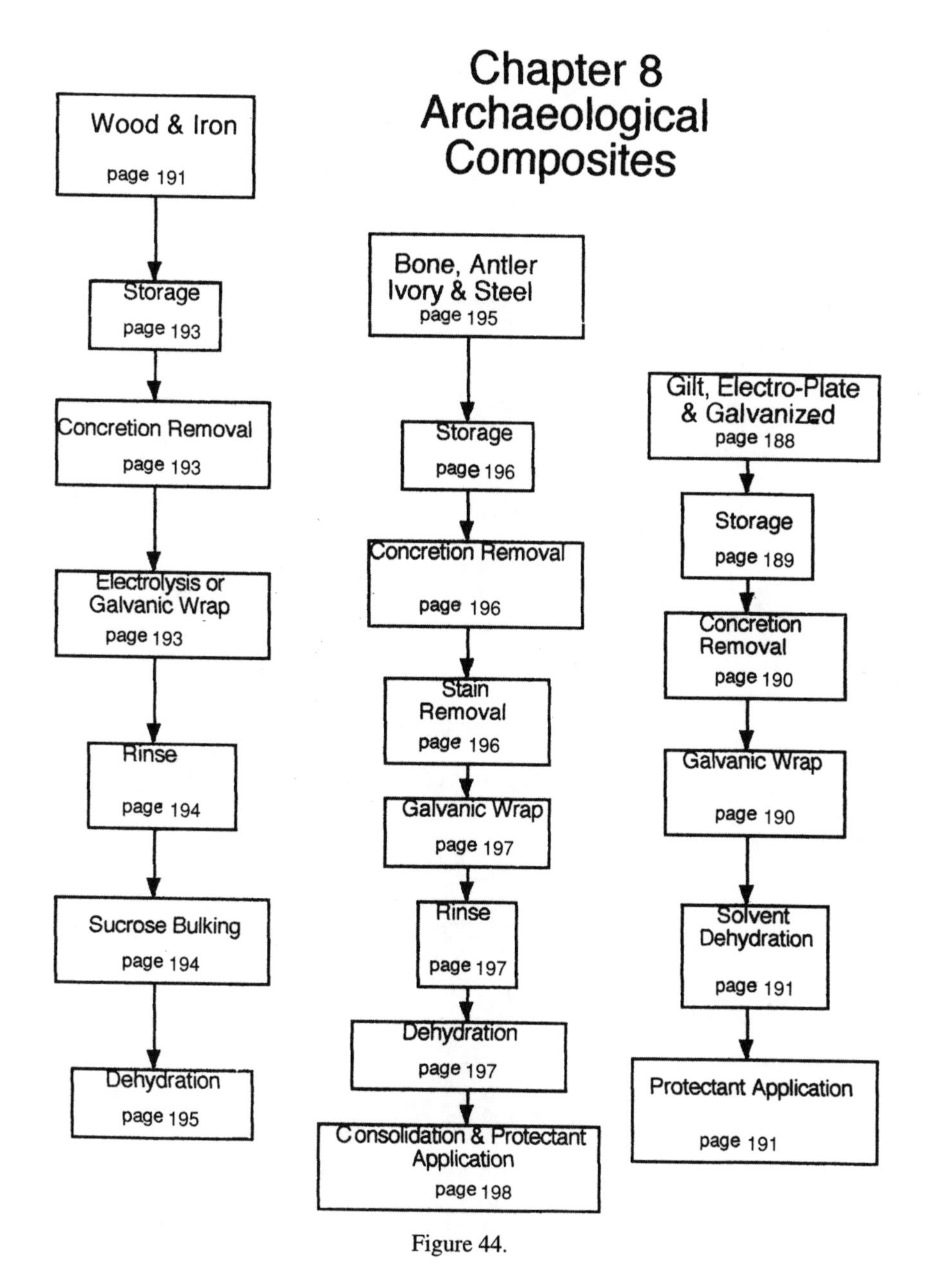
Chapter 8
Archaeological
Composites
Wood & Iron
page 191
Storage
page 193
Concretion Removal
page 193
Electrolysis or
Galvanic Wrap
page 193
Rinse
page 194
Sucrose Bulking
page 194
Dehydration
page 195
Bone, Antler
Ivory & Steel
page 195
Storage
page 196
Concretion Removal
page 196
Stain
Removal
page 196
Galvanic Wrap
page 197
Rinse
page 197
Dehydration
page 197
Consolidation & Protectant
Application
page 198
Gilt, Electro-Plate
& Galvanized
page 188
Storage
page 189
Concretion
Removal
page 190
Galvanic Wrap
page 190
Solvent
Dehydration
page 191
Protectant Application
page 191

Figure 44.

Chapter 8

Archaeological Composites

Composites as a group can be a conservator's worst nightmare. By definition a composite artifact is an object fabricated of more than one type of material. The problem with composites is that often the conservation procedure for one of the materials is damaging to the other. This is no problem if the artifact can be taken apart and the two materials treated independently. The best advice anyone can give concerning composites, is that they should be taken apart and the different materials stabilized and brought back together when they are finished. But this cannot always be so easily accomplished. Components of artifacts may in fact, be so closely attached or embedded that they cannot be separated without the utter destruction of the artifact. Fortunately non-separable composites are rare, and the problems are rarer still if the archaeologist/conservator uses good judgement and contacts other conservator's who have faced the same, or similar types of problems and were able to free the different materials for processing.

Included in this chapter are three of the most prevalent composite artifact groups that an archaeologist/conservator is likely to run up against. These include iron imbedded in wood; bone, antler, and ivory, connected to steel; and gilding or plating of one metal on another. Examples of the first group of artifacts include wooden pulleys and blocks with iron pins and strops as well as construction scantlings with through pins. The second group includes bone, antler, or ivory handled knives, forks, and spoons. Plating examples of the third group include tin cans and galvanized buckets that were hot plated by dipping them in molten tin and zinc, plus electrochemically plated silver, nickel, and gold watches, drawer pulls, handles, and jewelry. These artifacts were never intended to be disarticulated once they were manufactured and in fact, the manufacturers went to great lengths to make sure they did not come apart under normal circumstance.

Composite artifacts leave the archaeologsit/conservator with a dilemma, damage the artifact to take it apart, or attempt to treat it in one piece. In most instances, except for plating and gilding, the conservator should make an effort to disassemble these artifacts and treat the individual components. In the long run this will insure that the interfaces between component materials will not harbor contaminants and harmful compounds. It will also insure that the treatments intended to conserve one material do not alter or harm the other.

But experience has shown that some artifacts will be destroyed before they can be disassembled. Gilding and plating, of course, will last the lifetime of the artifact and should not be removed. But other multi-component objects may be fastened with metal pins that were peened, or held together with internal interlocking mortise and tenens. These types of artifacts will need special treatment and curation to insure their continued survival.

Composite artifact treatment is not always ideal, compromises are made to the individual components of an artifact, but these treatments will stabilized an object in one piece. Should a composite object show continued decomposition in storage or on display, more dynamic procedures may be called for. Conservation treatments can always be ramped up to meet a challenge, but physical destruction cannot be mitigated once it is done.

The following treatments offer an alternative to the demolition of an artifact and should be instituted after sincere attempts to separate components have failed. These methods are the best solution to a difficult circumstance.

GILDING, PLATING, AND GALVANIZING—THEORY

Chemically plated and electroplated buttons, metal ornaments, and jewelry have been produced since the middle of the 19th century. The process electrochemically plates a very thin layer of the more precious or noble metal on a base of iron, copper, copper alloy, or nickel. This plating can be as thin as 1/1000 of an inch or .002 cm. The advantage of this process is obvious for jewelry and ornaments as they could be made to look expensive when in fact they were only as costly as the base metal plus the plating process. Before electrochemical plating became popular for jewelry and ornaments, gilding achieved much the same goal as craftsmen attached thin sheets of silver and gold to less noble base metals or wooden objects by hand.

Tin cans and galvanized buckets came into use at about the same time as jewelry plating but for more utilitarian purposes. Nonetheless, cans are in fact, very sophisticated food storage devices that use the principle of galvanic coupling. The can itself is made of an iron alloy that is hot dipped into tin and originally was soldered together with lead solder. Since the tin is a more noble metal than iron it is reduced (receives electrons) from the iron and will remain in pristine condition both on the outside of the can and most importantly on the inside next to the preserved food. Unfortunately, tin was also more noble than lead in the early cans. Some of the lead in the solder that held the seams together corroded to preserve the tin and the lead ions migrated into the food. Before the problem was solved lead contaminated canned food created a major health hazard.

Galvanized buckets and other containers revolutionized the use of cheap iron alloys for utilitarian containers. Though cheap to produce iron alloy containers did not last long and easily corroded. Hot dipping the iron in the less noble metal zinc, solved this problem as the zinc corroded to preserve the object. Galvanized buckets eventually replaced leather and wood as the utilitarian material of choice.

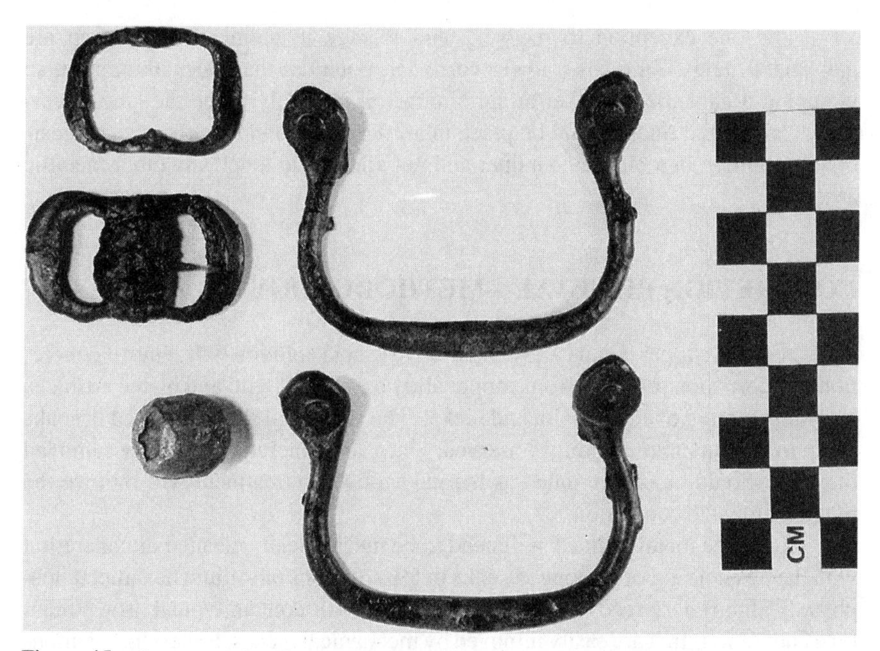

Figure 45. Some examples of composite metal artifacts—nickel plated iron drawer pulls and brass and iron buckles, along with a solid lead gaming piece. Photograph by author.

Gilding, plating, or galvanizing by definition places two metals together that have different reduction and corrosion potentials. This creates a galvanic couple in which the least noble base metal will begin donating electrons to the more noble plate or gilt. As electrons are donated, metallic ions are freed in the base metal. These ions will begin to migrate from the artifact and combine with other substances to form corrosion. Since the plating or gilt is being reduced it will survive quite well in the archeological record while the base metal will be less well preserved. The objective of the treatment for gilt and plated objects, therefore, is to reduce and conserve both metals at the same time.

STORAGE—METHODOLOGY

Recovered gilt and plated metallic artifacts should be recovered and stored as recommended with iron, copper, and miscellaneous metals Chapters 3, 4, and 5. Gilt and plated items including buttons, watches, and jewelry. All plated items should be wrapped in aluminum foil and placed in a 2% to 5% solution of sodium carbonate or bicarbonate for storage and transportation. Aluminum has a lower corrosion potential than plated or gilt artifacts and will begin to electrochemically reduce them in storage. The reduction is mild with virtually no gas generation and no worry that the plating will be damaged.

The one exception to recovery and storage in aluminum foil wrap are galvanized items. Zinc has a lower corrosion potential than does aluminum, so wrapping a galvanized bucket in aluminum foil will only cause the zinc to corrode. Galvanized finds should be place in a 2% to 5% sodium carbonate or bicarbonate solution in a plastic container and not allowed to touch any other metallic artifacts.

CONCRETION REMOVAL—METHODOLOGY

As with copper and its alloys a 10% citric acid solution will remove concretion and corrosion products from copper alloy base metal, gilt, and plated artifacts, including plating of silver, gold, and nickel. The citric acid should be used in soaks of up to 8 hours and carefully observed. Once the concretion is nearly removed the soak should be discontinued as the galvanic wrap treatment will remove the remainder of the concretion.

Iron base metal artifacts will need to be mechanically cleaned of concretion with the possible aid of prolonged soaks in 5% solutions of sodium hexametaphosphate (Calgon). Dry recovered cans and buckets will contain typical brown mass type concretions that are easily removed by mechanical means. Experience demonstrates that there is little chance that a can or galvanized bucket will survive long enough in an ocean environment to retain a concretion.

GALVANIC WRAP (ELECTROCHEMICAL CLEANING)—METHODOLOGY

As with the other metals covered in this text, the galvanic wrap treatment for gilt and plated artifacts on copper alloy base metals is a continuation of the recovery and storage process except that the sodium carbonate or bicarbonate storage solution should be replaced with citric acid. The citric acid produces faster and crisper results on the surface of the artifact. Copper and copper alloy base metal artifacts with gold, silver, or nickel plating should first be cleaned of excess concretion and dirt. A pouch of aluminum foil large enough to contain the artifact is prepared within another container (to contain the spillage) and filled with a 10% citric acid electrolyte. The artifact is placed in the pouch and it is closed and wrapped snuggly around the object, without damaging it. The galvanic wrap container should then be filled with more electrolyte to cover the foil pouch. Treatment times can take several days or last for a week or more. The slow gentle reduction of the artifact by the aluminum foil will generate very little gas or other harmful side effects.

Galvanic wraps for plated iron objects follow the same basic principles outlined for copper alloy base metals, except the electrolyte should continue to be sodium carbonate or sodium bicarbonate in a 2% to 5% mixture (a continuation of recovery and storage). Nickel and tin plated iron will be reduced gradually by this process with no gas generation or deleterious effects. Within days the iron and plate metal will be reduced and remaining concretion will fall away. If the artifact is heavily salt contaminated the electrolyte in the galvanic pouch will need frequent changes.

Galvanized artifacts will need to undergo a more active treatment than a galvanic wrap. Galvanized artifacts require the active intervention of electrolytic reduction, as the zinc coating is too easily oxidized in most rinses. Electrolysis is set up as it is for cast iron (see Chapter 3). Mild amperage and a light .25% to .5% sodium carbonate electrolyte will insure an adequate rinse of the artifact. Scrubbing afterward with nylon brushes will removed the dark magnetite coating on the outside of the artifact.

SOLVENT DEHYDRATION—METHODOLOGY

Heat treatment can be very detrimental to gilded objects. Therefore, solvent dehydration as described in Chapter 4 for copper and copper alloy artifacts should be used. Three successive one-hour baths in alcohol will suffice to dehydrate the object. After the third bath it can be removed and allowed to air dry.

PROTECTANT APPLICATION—METHODOLOGY

Protectant application is an option for many plated and gilded artifacts. Heat treatments should not be used to coat fragile or flaking objects. Micro-crystalline wax can be used for robust artifacts unaffected by the heat. Shellac is an old coating but a good alternative to wax. It is still effective and most importantly, reversible. The shellac is simply painted on with a brush or sprayed on in aerosol can application. Shellac is removed with any solvent and alcohol works well for this.

Incralac is a mixture of benzotriazol (BTA) in shellac. Though this formula has worked well in the past the conservator should be aware that BTA is a carcinogen and should only be used with gloves and a breathing mask in a well ventilated area. It is also a non-reversible treatment since the BTA will chemically bond with the copper alloy.

Wood and Iron

Since the early Iron Age man's favorite fastener for wood has been iron. Iron is plentiful and cheap, it is also one of the few readily available metals that is hard enough to drive into softwoods without a pilot hole. Wood and iron artifacts are by far the most plentiful and most difficult to treat objects belonging to the composite grouping. Separately, wood and iron are not difficult to treat and traditional conservation practice dictates that these artifacts be disassembled. Not surprisingly, however, composites are often difficult or impossible to disengage.

Combined wood/iron artifacts suffer accelerated decomposition as iron corrosion products degrade, stain, and mineralize the wood nearest the corroding object. This deterioration process makes the corroded or concreted iron difficult to remove without destroying much of the wood surrounding it. The iron corrosion products also tend to swell or enlarge from their metallic state making them fit even more snugly in the surrounding wood.

Dry recovered wood and iron composites can sometimes be taken apart by allowing them to dry in a low relative humidity of 40% or less. The wood should shrink at this humidity loosening the iron. Since dry recovered wood has already collapsed (if it was once waterlogged) the reduced humidity drying cannot hurt it further. Shrinkage from loss of water in the cell walls will correct itself as soon as the object is placed back into 50% or more relative humidity.

Oceanic and fresh water recovered artifacts are a different story entirely. Usually the wood in these artifacts is waterlogged and the seawater-recovered iron has absorbed, at least at its surface, a large concentration of chloride ions. Since waterlogged wood cannot be allowed to dry because it will collapse (see Chapter 2),

Figure 46. Wood and iron composite pulley block sheave. These artifacts are all but impossible to separate into their elemental components and can be treated as this was, with electro—sugar techniques. Photograph by Frank Cantelas.

there seems no solution to the problem of removing the iron. Especially since strong physical exertion may also damage the degraded wood.

One solution to this dilemma is to treat the composite in one piece. This is not easy, however, since the most commonly practiced conservation procedures for wood and iron are detrimental to one another. PEG activates corrosion in iron, and the strong bases used for some recommended electrolytic procedures at other laboratories would be injurious to wood, breaking it down at the cellular level.

The solution to this problem is a compromise that has produced good results in both esthetics and lasting quality of treatment. This approach uses low amperage light electrolysis suggested for metals in this manual (Chapters 3, 4 and 5), combined with sucrose bulking of the wood as suggested in Chapter 2. The most likely scenario in the wood/iron composite group is that the conservator will be forced to treat the protruding head of an iron drift bolt, rod, or fastener, while the rest of the iron remains untouchable inside the encasing wood.

STORAGE—METHODOLOGY

Wood/iron composites should be stored in fresh water after the salinity of the storage solution has been gradually decreased (see wood storage, Chapter 2). Under no circumstance should a strong base be placed in the storage solution to prevent the iron from corroding. Strong bases will disintegrate the interstitial wood cell bonding and break down the wood in the object. Exposed iron or other metals can, in this instance, be protected galvanically with a wrap of aluminum foil pressed tightly onto the metal surface.

CONCRETION REMOVAL—METHODOLOGY

All concretion removal from wood/iron composites should be done mechanically. Generally, concretions form near the iron but they can encase a good deal of the wood. In the case of the protruding drift bolt head a lump of concretion can form on it and cover the surrounding wood. While hammers and air scribes are used to remove the concretion liberal use should be made of protective padding and wetted cloth covering to protect the wooden part of the artifact. Since each artifact is unique, methods for transport and protection must evolve on a case by case basis. The aluminum foil wrap should be placed back on the projecting iron after it is fully exposed.

ELECTROLYSIS—METHODOLOGY

Chloride ion penetration of composites poses a major problem in the conservation of the metal component of wood/iron artifacts. Experiments and experience have demonstrated, that electrolytic reduction using dilute .25% sodium carbonate electrolyte at a pH just over 8 will not seriously damage the wood in the time it will take to rinse most of the chlorides from the metal surface.

The reduction is carried out as it is for low amperage reduction suggested in chapter 3. It should be remembered that electrolysis only works if good contact has been established between the direct current power source and the object being reduced. Since only a small portion of the iron artifact being reduced in this instance will be visible, it is imperative that it be well connected with the negative terminal of the power source. The positive terminal will be connected to a nearby mild steel sacrificial anode.

Reduction of some of the corrosion product to magnetite on the iron surface will foster the rinse process, allowing for chloride removal at the wood/iron interface. It is unclear how far the rinse process can go to remove deeply penetrated chlorides from inside the artifact. In some instances water has not penetrated to the core of an artifact so no chlorides are present. In other instances the waterlogging is complete and water and chlorides have penetrated even an imbedded iron object.

A general time frame for this process is that it will take longer to rinse the chlorides from an imbedded object than one that is being reduced on all of its surfaces. If a nail would normally take one or two weeks of reduction, an imbedded nail will take about four weeks. Even so, it is likely that only the head of the nail will get a thorough rinse, time will tell if the interior will need more treatment.

RINSE—METHODOLOGY

A careful and vigorous distilled or deionized water rinse should follow completion of the electrolytic chloride washing procedure. This will insure that little of the electrolyte remains on the artifact or has penetrated the wood. This rinse will take place in two phases. First a running water phase in which the iron is carefully scrubbed free of remaining concretion, corrosion product, and reduction residue. Then a static rinse phase in which the artifact is placed in deionized or distilled water for several days. The static rinse will allow any electrolyte that has penetrated the wood to rinse clear.

SUCROSE BULKING FOR THE WOOD/IRON COMPOSITE—METHODOLOGY

After the artifact is thoroughly rinsed it is ready to undergo wood conservation. Sucrose, the chosen bulking agent for this procedure, has demonstrated a mild temperament in regard to iron stabilization, unlike PEG which promotes corrosion. Experience has demonstrated that if anything, sucrose solutions seem to inhibit iron corrosion.

Bulking treatment of the wood follows the procedures outlined in chapter 2 of this manual. The artifact is submerged first in a 10% solution of sucrose and increase 10% on a weekly basis until a 50% solution is achieved. Aluminum foil wrap should be placed over the exposed iron during the bulking phase. Bulking treatments with sucrose should take place in heated tanks to insure success and to keep biotic growth to a minimal. Proper weight gain in the 20% to 35% range will signal the bulking agent's success at penetrating the wooden part of the object.

DEHYDRATION—METHODOLOGY

After bulking is complete the artifact is removed from the sucrose and excess solution is rinsed from the metal surfaces using distilled water atomizers or squirt bottles. The artifact should be slow dried in the relative humidity chamber described in chapter 2. When this procedure is complete the artifact should be stored in a fairly low relative humidity of between 40% and 50%.

The protruding iron can be sealed with a coating of tannic acid and shellac. This should help prevent water vapor and oxygen from being reduced at the metal's surface to the detriment of metallic iron inside the artifact.

Bone, Antler, or Ivory and Steel

Bone, antler, and ivory are formed of calcium salts built on a porous collagen matrix. The natural appearance of these materials is often smooth and hard and white or yellowish in color. Bone, antler, and ivory were at various times popular in jewelry, scrimshaw, inlays for furniture, and for utensil handles, particularly knives. In the archaeological record, bone, ivory, and antler are most often found in conjunction with steel blades.

Collagen handled knives are usually extremely delicate. Steel corrodes much faster than its parent metal iron and the acids produced by the corroding blade

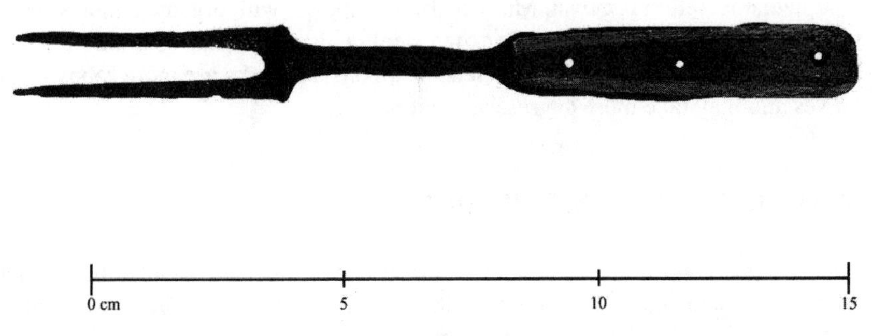

Figure 47. A bone handled fork treated as a composite artifact. Photograph by Frank Cantelas, digitally modified by Chris Valvano.

immediately attack the calcium salts in the handle. A bone handled steel knife is normally riveted in the handle, most often with brass rivets that cross through the tang that is sandwiched between the two cheeks of the handle pieces. The brass rivets exacerbate steel's corrosion qualities by setting up a galvanic couple in which the steel corrodes to protect the brass rivets. Unfortunately, the corrosion process usually permanently embeds the rivets insuring that the artifact cannot be taken apart without drilling the rivets. The corroding steel naturally stains the organic handle with insoluble iron salts and permeates acid into the organic material. Since the organic material is a natural base, they will eventually deteriorate while in contact with corroding metals.

STORAGE—METHODOLOGY

Dry recovered bone handled knives and utensils should be kept dry. It would not hurt to wrap the blade in aluminum foil. Wet recovered organic handled knives should be stored in fresh, distilled or deionized water to await treatment with the blades wrapped in aluminum foil. Under no circumstance should a base or alkaline solution be used to store the artifact. Most bases, even mild solutions of sodium carbonate or sodium bicarbonate will cause the collagen and calcium salts to swell and crack. The effects of the alkaline absorption are immediate and alarming.

CONCRETION REMOVAL—METHODOLOGY

Steel concretions, whether dry recovered or sea floor recovered should be mechanically or manually cleaned from the artifact. The steel will be quite brittle, possibly thin and weak as well, so this can become a real test of a conservator's patience. Dental picks and scalpels are the recommended tools, an air or electric scribe may set up too much vibration and crack or exfoliate the blade. During the concretion removal process the organic handle should be covered and protected from abrasion and shock. Light bead blasting of the steel may be necessary for complete concretion removal. Many forks and spoons with organic handles seem to have a base metal of iron rather than steel. This makes sense as they did not have to keep a sharp edge. For this reason spoons and forks are more robust than knives and may take more aggressive treatment.

STAIN REMOVAL—METHODOLOGY

At this point in the treatment all effort should be made to take the artifact apart and treat the individual components separately. Should this prove impossible without harm to the artifact the following treatment is recommended.

Organic handles are susceptible to damage from both bases and acids. Therefore, a 2% to 5% trisodium salt of EDTA solution with a nearly neutral pH, or a weak sodium hexametaphosphate solution in distilled water should be used to remove the insoluble iron salts from the blade handle. If the stains appear in only one area, a poultice application of the stain remover on cotton or talc can be used to deliver the chelating agent. If the entire handle is stained (as is usually the case), the knife is placed handle down in a beaker and the beaker filled to the top of the handle with the stain remover. The stain removal soak may last for several days but the artifact should not be left without frequent examination.

GALVANIC WRAP—METHODOLOGY

After the stains have been removed from the handle the chlorides can be removed from the blade. The galvanic wrap will be done as it is in Chapters 3, 4, and 5, except that only the blade will be treated. The blade should be wrapped in aluminum foil so that the electrolyte can infiltrate between the blade and the foil. The artifact should be inverted from its previous position so that the blade is now down in the beaker. Mild 2% sodium carbonate or bicarbonate solution is poured into the beaker to the top of the blade. Great care should be taken not to let the solution splash the handle as it will cause spot swelling. The solution should be changed each week and the treatment can continue for several weeks, depending on how badly the blade is saturated with salt.

RINSE—METHODOLOGY

After both the stain removal and the desalination wrap, an hour-long rinse in distilled or deionized water will remove the treatment chemicals. These rinses should follow the treatments themselves, meaning the stain removal rinse should only involve the handle, while the chloride and blade reduction rinse should follow the galvanic wrap.

DEHYDRATION—METHODOLOGY

Bone, antler, and ivory are sensitive to extremes in temperature, ruling out oven and freeze-drying. Solvent drying on the other hand may dissolve some of the natural oils found in the organic material, rendering the handle brittle and unnatural looking on drying. The handle, therefore, should be allowed to air dry while the blade should be solvent dried directly after its rinse to keep new corrosion from forming before the water dissipates. Solvent drying is the same for this composite as it is for copper, three successive baths of alcohol for one hour each. Obviously only the blade should be immersed in the solvent dehydration bath.

CONSOLIDATION AND PROTECTANT APPLICATION—METHODOLOGY

The blade and the handle are treated separately from stain removal onward, consolidation and protectant application is no exception. The blade could use a simple coat of protectant to ward off further moisture penetration. Shellac is easily applied by brush and simply allowed to air dry. It can be removed at any time with alcohol.

If, after treatment, the handle pieces are friable and flaking they may be consolidated with PVA based white glue emulsion (not all white glues are PVA based). Again, the artifact needs to be placed blade up and handle down in a beaker large enough to hold it. The glue can be mixed with distilled water in ratios of 10% to 50%. The lighter the glue percentage the more the penetration, the heavier the concentration of glue the greater the surface bonding. Several days soaking time should allow the consolidant to penetrate the handles. Afterward the handles should be allowed to slow dry under cover.

In the final analysis this process will work well for bone, antler, and ivory handled implements that are not too highly salt contaminated. The tang or the part of the utensil that lies sandwiched between the handle halves remains largely untreated and if the artifact shows signs of decomposition in storage it will first appear in the tang area. Finally, the artifact should be stored in a dark, non-acidic container, with about 50% relative humidity.

CONCLUSION

The composite artifacts mentioned in this chapter are a small reflection of the myriad combinations possible within the entire realm of archaeological materials. The great majority of these objects are most easily treated by separating them into their component materials, but this is not always possible. Textiles adhere readily to iron objects, cordage is often found in conjunction with wood, and leather shoes are often held together with iron nails. Fortunately prehistoric sites offer up very few composite artifacts due to the simplicity of early cultural material kits, but it is possible to locate hafted projectile points and leather wood combinations. Since artifacts are invariably unique they all offer particular conservation dilemmas, not so much beyond the scope of this work, but beyond effective description. It becomes pointless and inefficient after a time to describe all of the possible combinations. The archaeologst/conservator should be reminded that some of these combinations, no doubt, will be too difficult to deal with in a minimal intervention laboratory. But other artifacts are simply a matter of networking, imagination, knowledge, and experience away from stabilization.

CHAPTER 8: ARCHAEOLOGICAL COMPOSITES

Argyropoulos, V., C. Degrigny, and E. Guilminot. "Monitoring Treatments of Waterlogged Iron-Wood Composite Artifacts Using Hostacor IT-PEG 400 Solutions." *Studies in Conservation* Vol. 45 No. 4, 2000, pp. 253–264. [ECU-746]

Baker, Andrew J. "Corrosion of Nails in CCA- and ACA-treated Wood in Two Environments." *Forest Products Journal* 42:9 (1992): 39–41. [ECU-436]

Blackshaw, Susan M. "Comparison of Different Makes of PEG and Results on Corrosion Testing of Metals in PEG Solutions." *Maritime Monographs and Reports* (National Maritime Museum, Greenwich, U.K.) 16 (1975): 51–58. [ECU-254]

CCI Laboratory Staff. "Care of Machinery Artifacts Displayed or Stored Outside." *CCI Notes* (Canadian Conservation Institute) 15/2 (1993). [ECU-481]

Cook, C. "Tests of Resins for the Treatment of Composite Objects." *ICOM-WWWG Newsletter* (Wet Organic Archaeological Materials 14) [n.d.]: 35.

Cook, C., Dietrich, A., D.W. Grattan, and N. Adair. "Experiments with Aqueous Treatments for Waterlogged Wood-Metal Objects." *ICOM-WWWG Proceedings* (Grenoble) (1984): 147–159. [ECU-22]

Ems, Robert. "A Spanish Relic fran Saint Thomas Harbor." *American Rifleman* 126 (1978): 26–27. [ECU-118]

Gilberg, M., D. Grattan, and D. Rennie. "Treatment of Iron-Wood Composite Materials." *ICOM Proceedings* (Fremantle) (1987): 265–268. [ECU-209]

Green, Jeremy N. "The Armament from the *Batavia*: I. Two Composite Guns." *International Journal of Nautical Archaeology* 9:1 (1980): 43–51. [ECU-171]

Hawley, Janet K. "A Synopsis of Current Treatments for Waterlogged Wood and Metal Composite Objects." *ICOM Proceedings* (Fremantle) (1987): 223–243. [ECU-207]

Jack, E.J. and S.I. Smedley. "Electrochemical Study of the Corrosion of Metals in Contact with Perservative-Treated Wood." *Corrosion* 43/5 (1987). [ECU-489]

Logan, Judith A. "The Cross from Ferryland." *Newsletter* (Canadian Conservation Institute) (December 1987): 11. [ECU-52]

MacLeod, Ian Donald and Neil A North. "Conservation of a Composite Cannon *Batavia*, 1629." *International Journal of Nautical Archaeology* 11 (1982): 213–219.

MacLeod, Ian Donald, F.M. Fraser, and V.L. Richards. "The PEG-Water Solvent System: Effects of Composition on Extraction of Chloride and Iron from Wood and Concretion." *ICOM Proceedings* (Fremantle) (1987): 245–263. [ECU-208]

McClave, Ed. "Corrosion Related Problems." *WoodenBoat* 93 (1990): 94–113. [ECU-117]

O'Donnell, E.B. and Maureen M. Julian. "Identification and Conservation of an 18th Century British Hand Grenade." *Proceedings* (14th Conference on Underwater Archaeology) (1983): 112–113. [ECU-66]

—. "Conservation of a Disassembled British 18th Century Naval Hand Grenade." *Proceedings* (16th Conference on Underwater Archaeology) (1985): 80–81. [ECU-67]

Panel Discussion. "Composite Artifacts." *ICOM Proceedings* (Fremantle) (1987): 271–287. [ECU-295]

Rodgers, Bradley A. *The East Carolina University Conservator's Cookbook: A Methodological Approach to the Conservation of Water Soaked Artifacts*. Herbert R. Paschal Memorial Fund Publication, East Carolina University, Program in Maritime History and Underwater Research, 1992. [ECU-402]

Rodgers, Bradley A. and John O. Jensen. "The Electro/Sugar Conservation of a Waterlogged Wood-Iron Composite Artifact." (Unpublished Manuscript, East Carolina University, Program in Maritime History and Underwater Research) (1990): 1–18. [ECU-170]

Starling, Katherine. "The Conservation of Wet Metal/Organic Composite Archaeological Artifacts at the Museum of London." *ICOM Proceedings* (Fremantle) (1987): 215–221. [ECU-206]

Selwyn, L.S., D.A. Rennie-Bissaillion , and N.E. Binnie. "Metal Corrosion Rates in Aqueous Treatments for Waterlogged Wood-Metal Composites." *Studies in Conservation* Vol. 38 No. 3, 1993, pp. 180–197. [ECU-747]

Appendix

List of Journals [with some abbreviations]

Abbey Newsletter
American Antiquity
American Neptune
American Rifleman
American Society of Heating, Refrigerating, and
  Air-Conditioning Engineers . . . . [ASHRAE]
Archailogia
Archaeology
Archaeology Ireland
Australian Natural History
Aviation Equipment Maintenance
Bermuda Maritime Museum Quarterly . . . . [BMMQ]
Biblical Archaeology Review
British Archaeology Reports
British Corrosion Journal
Bulletin of the American Institute for Conservation
Bulletin of the Australian Institute for Maritime Conservation
Canadian Conservation Institute Newsletter . . . . [CCI Newsletter]
Canadian Conservation Institute Notes
Chemical Trade Journal and Chemical Engineer . . . . [CRJCE]
Chemistry in Australia
Chemistry in Britain
Christian Science Monitor
Conservation News
Conservation Science Bulletin
Conservator, The
Corrosion
Corrosion Australasia
Corrosion Protection and Control
Corrosion Science
Country Life
Curator

Education in Chemistry
Explorer, The
Florida Anthropologist, The
Forest Products Journal
Foundry Trade Journal
Gemnologist, The
Guild of Bookworkers Journal
Heritage Australia
Historical Archaeology
History News
International Counsel of Museums Proceedings........................[ICOM]
Industrial and Engineering Chemistry
International Journal of Nautical Archaeology.......................... [IJNA]
Irish Archaeological Research Forum
Institute of Nautical Archaeology Newsletter............................ [INA]
Journal of Applied Polymer Science
Journal of Archaeological Science
Journal of Chromatography
Journal of Coating Technology
Journal of Field Archaeology
Journal of Glass Studies
Journal of the Less Common Metals
Journal of the American Institute for Conservation..................... [JAIC]
Journal of the Association for Preservation Technology, The
Journal of the Canadian Conservation Institute..........................[CCI]
Journal of the Electrobiochemical Society
Journal of the Institute of Wood Science
Journal of the IIC—Canadian Group
Journal of the Japan Wood Research Society
Journal of the Oil and Colour Chemists Association
Journal of the Society of Leather Technologists and Chemists
Maritime Whales
Maritimes
Material Performance
Materials Protection and Performance
Medical and Biological Illustration
Metal Finishing
Metal Progress
Metallurgist and Materials Technologist
Metals and Materials
Museum
Museum News
Museums Journal
National Geographic

National Geographic Society Research Reports
Natural History
New Scientists
Newsletter of the National Museum of Antiquities of Scotland
Popular Science
Poseidon
Process Biochemistry
Realia
Science
Science and Archaeology
Science for Conservation
Science News
Science Technology
Scientific American
Scottish Society for Conservation and Restoration
Smithsonian
Society for Historical Archaeology Newsletter, The
Studies in Conservation . . . . [SIC]
Swedish Corrosion Institute Bulletin
Technology and Conservation Magazine
Technology and Culture
Textile Conservation Newsletter
Textile Month
Textile Research Journal
University of London Archaeological Bulletin
Vacuum
Water
Waterlogged Wood Working Group Newsletter . . . . [WWWG]
Washingtonian
Winterthur Newsletter, The
Waterlogged Organic Archaeological Materials Newsletter . . . . [WOAM]
Wood and Fiber
Wood Science
Wood Science and Technology
Wooden Boat
Yankee

Index